Henry Chamberlaine Russell

Observations of the Transit of Venus, 9 December, 1874

Made at Stations in New South Wales

Henry Chamberlaine Russell

Observations of the Transit of Venus, 9 December, 1874
Made at Stations in New South Wales

ISBN/EAN: 9783337186852

Printed in Europe, USA, Canada, Australia, Japan

Cover: Foto ©berggeist007 / pixelio.de

More available books at **www.hansebooks.com**

OBSERVATIONS

OF THE

TRANSIT OF VENUS,

9 DECEMBER, 1874;

MADE AT STATIONS IN NEW SOUTH WALES.

ILLUSTRATED WITH PHOTOGRAPHS AND DRAWINGS.

UNDER THE DIRECTION OF

H. C. RUSSELL, B.A., C.M.G., F.R.S., F.R.A.S., &c.,
GOVERNMENT ASTRONOMER.

Published by Authority of Her Majesty's Government in New South Wales.

Sydney:
CHARLES POTTER, GOVERNMENT PRINTER, PHILLIP-STREET.

5d 35—92   a   1892.

INDEX.

	PAGE.
Mr. Russell's report	1
Mr. Lenehan's report	8
Mr. Savage's report	10
Mr. P. F. Adams' report	11
Mr. Vessey's report	15
Mr. Hirst's report	19
Mr. Du Faur's report	20
Mr. Fairfax's report	23
Captain Hixson's report	24
Captain Onslow's report	27
Professor Liversidge's report	28
Mr. Tornaghi's report	29
Rev. A. Scott's report	30
Mr. MacDonnell's report	34
Mr. Watkins' report	36
Dr. Wright's report	37
Mr. Allerding's report	39
Mr. Bolding's report	39
Messrs. Belfield and Park's report	41
Mr. Belfield's report	42

TRANSIT OF VENUS, 1874.

LIST OF PLATES.

Nos. I to VII.—Mr. Vessey's observations, showing "Halo on disc," "Dusky ligament," "Ring of light," "Irregular shape of Planet," &c., &c.

Nos. VIII, IX, X.—Mr. Belfield's observations, showing "Illumination and colour of Planet at ingress and egress."

No. XI.—Mr. Savage's observations, showing "Flash of light on outer edge of Planet at egress.

No. XII.—Mr. Watkin's observations, showing "Elongation of Planet."

No. XIII.—Mr. Allerding's observations, showing the "Black drop."

Nos. XIV, XV.—Mr. Hirst's observations, showing the "Black drop" and "Fringe of red light."

No. XVI.—Captain Onslow's observations, showing "Halo on sun."

No. XVII.—Mr. Lenehan's observations, showing "Halo and polar spot."

No. XVIII.—Mr. MacDonnell's observations, showing "Halo around Planet."

Nos. XIX, XX, XXI.—Mr. Park's observations, showing "Positions of illuminated edges," "Blue colour of Planet," and "Illumination of preceding limb."

No. XXII.—Professor Liversidge's observations, showing "Illumination on Planet" and "Processes at egress."

No. XXIII.—Dr. Wright's observations, showing "Halo and polar spot at egress."

Nos. XXIV, XXV, XXVI, XXVII.—Mr. Russell's observations, showing "Vibration," "Halo," "Polar spot," "Haziness," &c., &c.

No. XXVIII.—Mr. Bolding's observations of "Ingress and egress."

No. XXIX.—Mr. Fairfax's observations, showing "Ring of Light."

No. XXX.—Mr. Russell's observations, showing "Ring of Light."

No. XXXI.—Sydney Observatory.

No. XXXII.—The 11¼-inch equatorial at Sydney.

No. XXXIII.—The photo.-heliograph.

No. XXXIV.—Transit of Venus, Camp, Woodford, New South Wales.

No. XXXV.—Waiting for the transit at Woodford.

No. XXXVI.—Waiting for the transit at Eden.

No. XXXVII.—The 7¼-inch equatorial used at Eden.

No. XXXVIII.—The 6-inch equatorial used at Goulburn.

No. XXXIX.—Photo taken at Sydney. Time 3h. 13m. 13·52s. S.M.T. (Reproduced directly from the negative.)

Inch squares placed at a distance of 400 feet and photographed with the Sydney equatorial as a test of distortion. (See page 2, Sydney observations.)

No. XL.—Photo. taken at Eden. Time, 1h. 9m. 55·89s., S.M.T. Photo. taken at Woodford. Time, 1h. 55m. 56s. S.M.T. (Both reproduced directly from the negatives.)

INTRODUCTION.

IMMEDIATELY after the very satisfactory observations of the transit of Venus, made by the observers in New South Wales, it was decided to

ERRATA.

Plate I, Fig. 3, the time should be 12 hours 15 minutes 30 seconds.
Plates VII and IX and X. *Read* Armidale, New South Wales.
Plate XII. *Read* Eden, New South Wales.
Plates XIV and XV. *Read* Mr. Hirst's observations at Woodford.
Plate XVII. Venus should be numbered 1, 2, 3, 4, beginning on the right.
Plate XVIII, should have Eden, New South Wales on it.
Plates XIX, XX and XXI. *Read* Armidale, New South Wales.
Plate XXII. *Read* Goulburn, New South Wales.
Plates XXIII, XXIV. *Read* Sydney, New South Wales.
Plate XXVIII, on diagrams 4 and 5, *read* "egress" for "ingress," "Bolding" for "Dalding."
Plate XXIX, for Sydney, New South Wales, read Woodford, New South Wales.
Plate XXX. *Read* ring of light at ingress, an enlarged view of fig. 3, plate XXVII.

Corrections to Mr. Vessey's Report.

	h. m. s.	h. m. s.
Page 16.	For 12 7 00·00	read 12 7 40·00.
Page 17.	„ 3 46 0·00	„ 3 48 0·00.
Page 18.	„ 12 23 51 00	„ 4 23 52·00.

Astronomer for New South Wales, died, having taken no steps to prepare for the transit of Venus observations. He had, indeed, expressed his intention of taking no part in the work, owing to other pressing duties. My appointment followed immediately after his death, and I at once took steps to prepare for the great astronomical event, fully realising the great importance of taking advantage of our favourable geographical position on the eastern coast of Australia for observing the egress. It was obviously for the honor of the Colony, as well as for the advancement of science, that the observations and photographs of the transit of Venus should be as complete as possible. I therefore at once brought the matter under the notice of the Government, asking authority to carry out the work, and a vote to cover the expenses. My request was warmly supported by our

INTRODUCTION.

IMMEDIATELY after the very satisfactory observations of the transit of Venus, made by the observers in New South Wales, it was decided to publish in full the result of their labours with coloured plates and every detail carefully revised by each observer, so that the printed diagrams should represent in each case exactly what he meant by his drawings; but the observers lived in widely separated districts, and so much time was lost in revising the proofs that a considerable period elapsed before they were ready. Meantime the Royal Astronomical Society deemed the reports and drawings of sufficient importance to publish them, and had published and circulated them before this volume was ready, and although some additional matter and diagrams will be found herein, it was at the time deemed better not to publish at once. Various circumstances, however, combined to delay the publication much longer than was at first intended; but it is hoped that the work will still be valuable as a record of the New South Wales observations for the present and probably the next generation of transit of Venus observers.

A short account of the circumstances which led up to the observations of the transit of Venus in New South Wales, will furnish explanation of various matters which need such reference, notably the delay in commencing the preparatory work, the instruments, and other apparatus used.

In August, 1870, Mr. G. R. Smalley, who had been the Government Astronomer for New South Wales, died, having taken no steps to prepare for the transit of Venus observations. He had, indeed, expressed his intention of taking no part in the work, owing to other pressing duties. My appointment followed immediately after his death, and I at once took steps to prepare for the great astronomical event, fully realising the great importance of taking advantage of our favourable geographical position on the eastern coast of Australia for observing the egress. It was obviously for the honor of the Colony, as well as for the advancement of science, that the observations and photographs of the transit of Venus should be as complete as possible. I therefore at once brought the matter under the notice of the Government, asking authority to carry out the work, and a vote to cover the expenses. My request was warmly supported by our

Royal Society, who appointed a deputation to wait on the Government to urge the importance of this work. Eventually, the Government gave authority, and a vote of £1,000 to enable me to carry out the work. This was in 1872, and inquiry elicited the fact that instrument-makers in Europe had their shops full of work for European observers, and could not undertake my orders at such a late period. Eventually, I succeeded in getting one photo-heliograph with Janssen apparatus, and also the object glass and micrometer for the large equatorial. For other apparatus required, I had to look to mechanics in the Colony, for the most part unused to such delicate work. This meant a great deal of additional work in designing, overlooking, and generally seeing after what was done. To provide for the photographic work, three equatorials were fitted up with photographic apparatus with which to take pictures of the sun; others were mounted on rigid stands; chronographs, buildings, &c., had to be made necessary to equip four observing-parties, each having transit instrument equatorials, an independent observatory and apparatus for observing and photographing the transit. By unremitting exertions everything was ready by the middle of 1874, and the work of organising and training the observers began, each party working in its own observatory, and with the apparatus they were going to use at the several observing stations, but which, for convenience, had been set up in the observatory grounds. Everything being quite ready the observing parties started in good time for their several localities. The reports which follow record their success. It would have been quite impossible to accomplish this had I not received from all who could assist most enthusiastic support.

The following *résumé* of the results was written immediately after the work, while the written and verbal reports of all the observers were fresh in my mind. It will serve as an introduction to what follows, and to call attention to some points of importance bearing upon the physical aspect of the ingress and egress of the planet, which may be useful to future observers, and, perhaps, in rediscussion of the transit of Venus, values of the parallax. But since these following observations were incorporated in the final results obtained by British observers, and fully discussed by the Astronomer Royal of England, Sir G. B. Airy, and also under his direction by Captain Tupman, I do not propose to enter into that question here, further than the quotation of a few sentences from the report upon the general result of all the British observation, which was prepared by Captain Tupman.

INTRODUCTION.

There were thirty-one British observers of ingress, whose reports were available. Of these eight were from New South Wales; but after weighting and discussion, the total number was reduced to twenty. Of these eight received double weight; and of these double-weighted eight, two, Messrs. Scott and Russell, are New South Wales observers.

At egress there were forty-eight observers, and again New South Wales furnished eight. After weighting and discussion, forty-eight are reduced to forty-one, but all the New South Wales eight remain, Vessey and Russell getting double weight, and Allerding, Captain Hixson, Lenehan, Liversidge, Dr. Wright, and Tornaghi, single weight. Captain Tupman, on page 453 of the monthly notices, R.A.S., June, 1878, says:—" It is seen that the observations at Sydney, New South Wales, viz., by Russell, Lenehan, Wright, and Allerding, have great weight in lowering the value of the parallax, and this effect is exaggerated by allowing Russell double weight. Mr. Russell's is one of the most detailed observations made; it seems impossible, from his description, to choose his earlier time. Mr. Lenehan admits he was late, and I have taken 10 seconds from his time, but still it is late. Dr. Wright made 'the most accurate observation possible', and agrees to a fraction of a second with Mr. Russell. Mr. Allerding does not profess to have made a good observation. Out of mere curiosity I have solved the thirty-five equations that remain, when Wilson's, Strahan's, and those at Sydney are excluded, and find the mean solar parallax as $8''\cdot 894$." In the previous discussion he made the parallax at ingress, in which the New South Wales observations were retained, $8''\cdot 845$, and at egress $8''\cdot 846$, the mean of which, $8\cdot 8455$, he accepted as the value of parallax from 1874 transit of Venus observations. Subsequent discussions show that this was a wise course, for the most elaborate discussion of the solar parallax which has yet been made, viz., that by Professor Harkness, combining the results from all available methods and published this year, 1892, makes the mean of all the values obtained since 1769 to be $8\cdot 834$, which shows clearly that Captain Tupman was right in retaining the New South Wales observations, although at the time they seemed to make the parallax smaller than it was thought to be. This is strong confirmation of what appears in subsequent pages, viz., the accuracy of the observations, made so by the very favourable state of the atmosphere, and good telescopes.

As already stated, it is not proposed to go into a discussion of the results here. I have merely quoted Captain Tupman to show what he thought of New South Wales observations, and their effect upon the value of the solar parallax, derived from 1874 transit Venus, which was, as we have seen, to make it much smaller than was at that time supposed to be its value, and to bring it more into accordance with all that astronomers have learned about it up to the present year, 1892. It will be seen in the following reports (page 5 and others) that the effect of bad definition is to make the observer think contact has been made before it really is made, and therefore a steady atmosphere is a most important factor in the value of such observations, any unsteadiness of atmosphere tending to make contact appear too soon, and therefore to make the solar parallax too large ; turning now to the *résumé* prepared at the time.

Never perhaps, in the world's history, did morning dawn on so many waiting astronomers as it did on the 9th of December, 1874. They were all anxiously looking for an answer to the old question, *to be, or not to be*, and certainly none could have expected a finer day than that which dawned on the observers of New South Wales. From all stations, in return for the morning clock signals, came the welcome intelligence that the morning gave promise of a splendid day, and after hearty good wishes had been given and received, we all turned to the final touches, which were necessary to complete our arrangements, and when these were done, waited, not without an involuntary feeling, which I will not call excitement, for that by common consent had been banished, but rather an overpowering sense of that responsibility which every true worshipper of science must feel, when he knows that the answer to half a century's questionings is depending upon him ; and that he is the observed of all observers ; but each one was determined to do his best in the noble cause of science ; supported by a faint hope that his name and his work would appear ages hence in the records of science, and be criticised under that blaze of knowledge which the united efforts of the world's science should produce.

And here it may not be out of place to introduce a few words about the selection of the New South Wales stations. For ingress there was little choice, for, the sun being in the zenith of a place near the longitude of Sydney, and in 23° south latitude, the parallax was almost nothing everywhere. At egress, however, our circumstances were much improved in this respect, and the south-eastern point of New South Wales was one of the best points of observation in Australia. In addition to this, there were two conditions to be borne in mind in making the selection—viz., weather, and telegraphic convenience for determining longitude.

I had for two years previously caused special meteorological observations to be taken at various places during the month of December. These indicated Woodford as about the most promising station for clear and steady atmosphere, and made it evident that the observers should not all be stationed at Eden, but be divided between the coast and the mountains. Bathurst and Goulburn were alike in chance of clear weather; but Goulburn was the better station geographically, and was therefore selected. Eden weather reports were not encouraging; but as the advantage of the position was so much in its favour, it was decided to make it the fourth station, and the Rev. W. Scott, formerly astronomer for New South Wales, proceeded there with Messrs. W. J. MacDonnell and J. S. Watkins, observers; and Mr. Sharkey, photographer to the Government Printing Office, as photographer, and one carpenter, with observatory, tents, instruments, and all needful appliances. The telescope used by Mr. Scott was a $7\frac{1}{4}$-inch equatorial of 10 feet 4 inches focus (*see* plates XXXVI and XXXVII); by Mr. MacDonnell, $4\frac{1}{4}$-inch Cooke equatorial; and Mr. Watkins, a $3\frac{1}{2}$-inch equatorial; they had also means of taking 220 photographs.

Captain Hixson, President of the Marine Board, Captain Onslow, M.P., and Professor Liversidge, with Mr. Tornaghi, photographer, and a carpenter, made the observing party at Goulburn; they took with them the 6-inch equatorial with the camera as shown in plate XXXVIII, the observatory tents; instruments, &c., which were similar to those at Eden, but the telescopes were smaller, having 6 inches, $3\frac{3}{8}$, and $3\frac{1}{4}$ inches object glasses. They had also means of taking 220 photographs.

P. F. Adams, Esq., Surveyor-General, with Messrs. Hirst, a well-known amateur astronomer, Mr. Vessy, of the Trigonometrical Survey, Mr. Du Faur of the Survey Department, Mr. Bischoff, photographer, and two carpenters, proceeded to "Woodford" the mountain residence of A. Fairfax, Esq. Their instruments were the photo-heliograph and Janssen apparatus, a $4\frac{1}{2}$ inch equatorial and one of $3\frac{1}{2}$ inches, chronometers, clock-chronographs and all necessary apparatus for 220 photographs, and 30 Janssen plates, each to hold 60 pictures. The observatory was similar to that used at Goulburn. (*See* plates XXXIII, XXXIV, and XXXV.) Mr. Adams gave me every assistance, and provided Transit Instrument and another observing telescope, and two ordinary tents.

At Sydney the observations depended wholly on the observatory staff. The instruments were the $11\frac{1}{4}$ equatorial (*see* plate XXXII), a $4\frac{3}{4}$ Equatorial by Troughton and Simms, and a 10-inch Browning-With silvered glass reflector, kindly lent for the occasion by J. Usher Collyer, Esq., with all usual clocks, chronometers, chronographs, &c., and means of taking 220 photographs.

I cannot leave this part of my subject without expressing my warm thanks to all who assisted me in observing the transit of Venus. With a zeal worthy of the occasion, one and all devoted themselves to a course of previous practice at the Observatory which involved an amount of hard work and self-denial worthy of all praise. This in some cases extended over several months in order to make themselves thoroughly conversant with their work; and this not only during daylight working at the artificial transit and photography, but also at night adjusting the instruments. My thanks, however warmly given, can be no measure of their work. I shall always feel grateful to them for their support, and for the enthusiasm and the zeal with which they devoted themselves to the work, and for their generous and thoughtful kindness and assistance to me personally in a thousand ways. From the officers of the Telegraph and Railway Departments also we received the most cordial assistance; indeed everyone for once made common cause with the Astronomers.

Previous to starting, all the observatories and instruments were set up in Sydney, and each party went to work in their own observatory; this we found to be of great service in pointing out weak points, which required either more practice, or the instrument maker to set right.

For practice in observing, we had two artificial transits, one similar to that designed by the Astronomer Royal, the other constructed in the Colony. The one made in Sydney consisted of first a sheet of metal, out of which a hole was cut 5 inches wide and 18 inches long, bounded on three sides by straight lines, and on the fourth by a curve of such radius that at a distance of 400 feet from where the telescopes were placed, it appeared like a section of the sun's limb. Behind this opening a piece of ground

INTRODUCTION.

glass, worked in a slide, and on it was fixed a blackened disc of metal having the apparent diameter of Venus. This was drawn along by clock-work, so that the artificial Venus appeared to come on to the sun's limb at external contact, and gradually travel in and make internal contact, at which point a most satisfactory black drop was seen. A large mirror behind all to reflect a bright part of the sky completed the apparatus. Five telescopes were directed to this, and as many observers, each using his chronometer, observed ingress and black drop, and then compared their observations. With this a great deal of practice was obtained, which was useful in training for observation. At the same time all were warned that there was no certainty about the black drop phenomena.

We come now to the day's work, and take first the weather at each station.

At Eden, the morning was fine and very promising, but about 11 a.m. clouds began to come, with a fresh sea-breeze, and led the observers to anticipate a disappointment. Fortunately up to the time of ingress the clouds had not interfered with the observations; but from that time forward the sun was more or less obscured, and at one period wholly so for 80 minutes, so that few photographs could be obtained, and the sun was entirely obscured some time before egress.

At Goulburn, the morning was fine, with light westerly wind and a few drifting clouds; during the afternoon the wind increased to half a gale, and the clouds were more numerous, but not sufficient to interfere with observation.

At Woodford, the morning was fine, with a dry hot wind (westerly), which increased as the day wore on; during the afternoon a few clouds passed over and interrupted the photographic work for a short time, but at ingress and egress the weather was splendid for observation.

At Sydney, the early morning was beautifully clear until 5h. 30m. a.m., when a heavy bank of fog came in from the sea and obscured the sun for three hours; but we still expected a fine day, and were not disappointed, for by 9 a.m. we had a clear bright sky and light north-easterly wind, which increased to a fresh breeze during the afternoon. The state of the atmosphere also was favourable for observation until the transit was over, except a few moments of bad definition; but had we been one hour later,

I do not think observations of egress would have been worth anything, for clouds were rapidly forming in the S.W., and, though thin, they spread very quickly over the sky.

Ingress.

For the purpose of convenient comparison I have arranged the times of observation in a tabular form, and put all into Sydney time. I confess when I saw the gradual phenomena of the transit myself I did not expect such a satisfactory agreement between the times of observation as some of the results show ; and if $4\frac{1}{4}$ seconds be taken as a fair estimate of the probable uncertainty of observed time at one station, when a definite phenomenon like the breaking of the black drop had to be observed, I think we, with no such definite phenomenon to observe, may congratulate ourselves that the differences are in most cases so small.

Only three out of thirteen observers took the time of first external contact, and they were evidently a few seconds, probably about ten, late. Such at least was my own impression at the time, for Venus had made something more than contact,—it was a small notch in the sun's limb.

My time is	... 11h. 55m. 23·00s.
Mr. Lenehan	... 11h. 55m. 30·34s.
Woodford—Mr. Vessey ...	... 11h. 55m. 14·96s.

These times show a very satisfactory agreement, especially when the difficulty of seeing external contacts is taken into account, and the fact that in 10 seconds of time Venus would only encroach about half a second of arc on the sun's limb, a quantity not easily seen. For second contact I think there can be no doubt that different phases of the phenomenon were taken by observers according to the different effects produced by the ghost of the black drop, which up to that time had a very tangible existence for all of us, not only from what we had read about it, but from seeing it so constantly in the artificial transit ; and as it is very important that the exact phenomenon taken by each observer as internal contact should be on record, I will here quote from the reports, beginning at Eden.

Mr. Scott took the time when 'he saw the partial obscuration of the sun's limb by the planet's atmosphere gradually diminishing until it disappeared altogether at 12h. 24m. 48s.', which I take to mean the completion of the sun's outline, the same phase which, as will be presently seen, I and others took for complete ingress.

Mr. MacDonnell took the time when 'the light seemed to be going in and out several times and prevented any accurate determination of complete ingress as 12h. 25m. 14·7s., but he is convinced he was 15s. late, making the time 12h. 24m. 59·70s.'

At Woodford, Mr. Vessey took the time when Venus appeared to touch the sun's limb, or when the two limbs were tangential. (*See* Plate II.) Time, 12h. 23m. 47·07s.

My own report of this phase is as follows, and it will observed that the first time given is 4 minutes before contact.

At 12h. 20m. 0s. indications of distortion or bad definition of the limbs in contact appeared, like a mass of black wool laid over the place, rendering it impossible to see distinctly and making the cusps very hazy. (See Plate XXVI.) I thought the drop was going to form, and watched very closely for it and for apparent contact, but I found it extremely difficult to make up my mind about the latter, and saw nothing of the former. 12h. 20m. 51s. was noted as a very unsatisfactory apparent contact. The cusps after this appeared to clear up or improve in definition (the telescope had not been altered), and as they approached each other the sharpness was very remarkable, but the motion so gradual, that I could not determine to a fraction of a second when they actually formed the line of light which I saw complete and took for the moment of internal contact, but the instant I was sure I made the record on the chronograph, which was at 12h. 23m. 59s., and keeping my eye steadily upon it saw it had in fifteen seconds become an unmistakable band of sunlight.

Mr. Lenehan says at time of ingress there was an indistinct shading between the supposed edge of the planet and the sun, which for some ten or fifteen seconds before the time I quote later, kept me in a state of uncertainty as to the true time of actul ingress; the shading did not break abruptly, but seemed to melt away in such a manner as to leave a doubt in my mind of the exact time the planet passed the edge of the sun, but I distinctly saw a clear band of light at 12h. 24m. 48·34s.

Mr. Savage says: 'The definition at this point being so very bad between the limbs of the sun and planet, and the edges at contact so very dark as to defy accuracy, as the planet advanced on the sun a little way this shading still connected the planet with the sun's edge, but that portion

INTRODUCTION.

of it nearest to the planet showed indications of fading away gradually, until at length it disappeared altogether without any sudden break whatever, and at 12h. 23m. 43·93s. a streak of light became visible between the planet and the sun's limb.'

Dr. Wright noted '12h. 24m. 30s., but was quite sure this was late, probably 30s., making 12h. 24m. 0s., having lost true contact looking for the black drop.' These times are :—

	h.	m.	s.
	12	24	48·00
	12	24	59·70
	12	23	47·07
	12	23	59·00
	12	24	18·34
	12	23	43·93
	12	24	0·00
Mean	12	24	18·00

But I think it is evident that only Mr. Scott, Mr. Vessey, and myself have taken exactly the same phase here, and the mean of the three results is 12h. 23m. 57·12s.; the differences are too great to give a satisfactory mean from all the observers, but if taken it is 12h. 24m. 18·00s. Perhaps some of the differences may be attributed to differences of temperament.

EGRESS.

For the third and really most important phase we had all fortunately learned to disbelieve *in black drops*, and during the photographic work had time to think and talk over what had been seen at ingress, and we went to our telescopes much better prepared for the work before us; still the difficulties were by no means gone, and the motion of the planet was so exceedingly slow that a few seconds variation is, I think, a necessity.

It was unfortunately cloudy at Eden, but the Goulburn observations now make up for it.

Captain Hickson saw internal contact	3h. 54m. 28·01s.
Professor Liversidge	3h. 54m. 20·37s.
Mr. Tornaghi	3h. 54m. 25·79s.
At Woodford, Mr. Vessey says	3h. 54m. 37·50s.

The circles of sun and planet tangential, and the ring of light about its own thickness outside the limbs of the sun.

My own time for this phase is ... 3h. 54m. 39·66s.

After a period of bad definition my report says, 'the limbs recovered their perfect definition and were clearly and steadily separated by a line of light which at 3h. 54m. 26·3s. could not have been more than half a second of arc in thickness, and then the same marvellous definition continuing just when it was wanted. The line gradually contracted to a scarcely visible thread, and the limbs made contact. There was no sudden break, nothing but the perfectly gradual motion of the one disc over the other, both beautifully defined, and I saw one overtake the other at 3h. 54m. 39·66s.'

Mr. Lenehan says the first apparent contact was at 3h. 54m. 21·61s., a little jumping; afterwards, saw a band or faint and narrow streak of light between the limbs of planet and sun, which clearly showed me that the time above given was too soon. I then waited until I was absolutely certain contact was complete, at 3h. 54m. 46·61s., but I feel confident this time is from 7s. to 10s. after true time, making the true time ... 3h. 54m. 39·61s.

Dr. Wright makes time of contact ... ... 3h. 54m. 39·59s.

Mr. Allerding makes time of contact ... 3h. 54m. 35·00s.

Now at Goulburn at this time the wind had become very strong, and produced a tremulous motion which would no doubt account for the times being a little early, for it would not be possible under those conditions to see a very fine thread of light, and we know that such was seen by the observers who agree best. Mr. Lenehan's time also is known to be late, and we have five times left of which the extreme difference is 4·66s., and mean 3h. 54m. 38·27s.; and two at Sydney—my own and Dr. Wright's—agree within less than one-tenth of a second.

For last contact we have only three observations which do not accord very well. At Woodford, Mr. Vessey took the time as ... ... 4h. 23m. 52·00s. and says : "This observation appeared to be correct to a small fraction of a second. The indentation on sun's limb gradually contracted in width till within thirteen seconds of time given, and it then seemed to contract longitudinally till it became a small notch like a boiling indentation. This was seen steadily diminishing till it suddenly flashed out, and the limb of the sun became perfect."

My own observation makes this ... ... 4h. 24m. 27·00s. At this time the last sign of the planet on the sun's disc was seen as the faintest possible mark, which then disappeared, definition being for the time very good, and the observation quite satisfactory.

Mr. Lenehan saw last and final contact at ... 4h. 23m. 49·61s. the edge of the planet being then lost in the edge of the sun.

My own observation of this phase does not seem to be supported, but the larger aperture of the telescope I used, the 11¼ inch reduced to 6 inches, and the steady motion by clock-work, probably explain the difference.

Turning, now, to the physical phenomena observed, there are several of them very interesting and important that will repay a little consideration, and first in regard to the *black drop* so called. The account of this phenomenon given by Mr. Stone, Astronomer Royal at the Cape of Good Hope, seemed so thoroughly satisfactory that I fully accepted it, and in common with nearly all observers, expected to see the planet distorted into a pear shape as it left the sun's edge, "as if a stalk or ligament connected it with the sun's limb" (*see* Mr. Hirst's observations) which broke suddenly ; a phenomenon the exact time of which could have been easily determined, but instead of this a set of wholly unexpected phenomena presented themselves.

As the planet encroached on the sun the cusps remained perfectly sharp until near the time of contact of the limbs, when a curious hazy appearance became developed, and rendered it impossible, in spite of all efforts, to see exactly what was going on. Most, if not all the observers, thought the *drop* was forming, but close attention only revealed a gradual disappearance of the haze until the sun's and planet's limbs were left perfectly clear and sharply defined with a thread of sunlight between them.

Rev. W. Scott says, in reference to this point:—'I continued to watch the planet for more than three minutes, and saw the partial obscuration of the sun's limb by the planet's atmosphere gradually diminishing until it disappeared altogether.'

Mr. MacDonnell says:—' As Venus proceeded the shadowy envelope disappeared, except between the planet and the sun's limb, where it seemed to fill up the space between them with faint rings concentric with the planet's edge. There was no distinct rupture of this appearance, the light seeming to go in and out several times.' Professor Liversidge says, 'A faint hazy gray filament like a streak of smoke was momentarily observed between the edge of the planet and the sun; it was very obscure and illdefined.'

My own report for ingress is as follows, at 12h. 20m. indications of distortion or bad definition of the limbs in contact appeared, like a mass of black wool laid over the place, rendering it impossible to see distinctly, and making the cusps very hazy. I thought the drop was going to form, and watched very closely for it, and for apparent contact, but I found it extremely difficult to make up my mind about the latter, and saw nothing of the former. 12 hours 20m. 51s. was noted as a very unsatisfactory *apparent contact*. The cusps after this appeared to clear up or improve in definition, and as they approached each other the sharpness was very remarkable.

At egress a curious phenomenon then presented itself similar to that remarked at ingress; the two limbs at the point of contact seemed to get confused or badly defined, whether from atmospheric causes near us, or some peculiarity about Venus, I am unable to say, but it seemed to disturb the planet in a most remarkable way.

Mr. Lenehan says, at the time of ingress there was an indistinct shading between the edge of the planet and the sun, which for some seconds kept me in a state of uncertainty as to the true time of actual ingress; the shading did not break abruptly, but seemed to melt away in such a manner as to leave a doubt in my mind of the exact time the planet passed the edge of the sun.

It is evident that what we have here described is a phenomenon very different from that which is known as the *black drop*, for here the uncertainty lasts much longer, and does not occur when the limbs are apparently separated, but when they are in fact, as well as appearance, in contact, and slightly overlapping, and while this phenomenon is clearly made out to have lasted about four minutes. Mr. Stone, Astronomer-Royal at the Cape of Good Hope, and the best authority on this subject, estimated that the black drop would only last 18 seconds.

Of the drop phenomena which we all expected to see we have two particularly interesting accounts, which I will quote. The first is that by Mr. Hirst, who was thoroughly acquainted with the phenomenon as described by Mr. Stone and others, and had practised with the artificial transit, though the work which was specially his, and of which he had made himself master, was the management of the photo-heliograph during the taking of the Janssen pictures.

Attached to the tube of the photo-heliograph was a finder consisting of a *single* non-achromatic lens $1\frac{1}{2}$ inch aperture and 4 feet focal length. This was originally arranged by the maker so as to throw the sun's image on to a piece of parchment fixed at its focus; but in order to adapt it to circumstances which required that one end of the heliograph should be in the photographer's dark room, the lens was inserted in the end of a brass tube, an eye-piece being provided in the shape of a Huyghenian combination, giving a power of about 50 diameters. The chromatic and spherical aberration of the single lens were not obtrusive owing to its extreme focal length, so that fair definition could be obtained of the edge of the sun, and the existence of even minute solar spots made plainly visible.

Of the drop as seen with this, Mr. Hirst says:—'To diminish the light in the finder I used a thick piece of orange-coloured glass, which gave an agreeable image of the sun. This was placed outside the eye lens of the eye-piece.

I had prepared and placed a plate in the Janssen apparatus, when, on taking my usual glance at the finder, to see that the telescope was adjusted ready to take photographs, I observed the disc of Venus appearing, as it were, rather more than one-third her own diameter within the sun, and connected with the limb by a narrow line intensely black, with an ill-defined edge. Plate XV represents the appearance as faithfully as I can recollect; this was about five seconds before No. 5 Janssen plate was begun. I had not time for more than a glance, for I wished to procure a photograph of what I supposed to be the black drop, so universally observed by astronomers, more than a century ago, at the last transit. On getting the plate through, however, it showed nothing of what I had so distinctly observed a few seconds before.

Referring to the finder, Venus appeared well inside the sun, but apparently *nearer* the limb than she seemed before. The drop was gone. I thought at the time that it might have broken before the exposure of the plate, and I determined to keep a sharp look-out for its formation at egress. Soon afterwards Mr. Vessey came in and reported that the 4½-inch had shown no drop at all.

Towards egress I referred constantly to the finder, that I might be ready with a plate directly the drop became visible. When Janssen plate No. 9 was in its place, and upon adjusting with the finder, I observed no black drop, the planet appearing so far within the sun's disc that I did not think it necessary to hurry in order to catch the drop and exposed the No. 9 plate, meaning to get another in time. After taking out the plate, which probably occupied twenty seconds, I went to the finder, and to my astonishment saw that the drop had formed, appearing about as long as one-third the diameter of the planet. I hurried on the next plate as much as possible, but a delay unfortunately of a couple of minutes occurred before it was ready; on development it showed Venus a perfectly circular disc touching the sun's limb. It appears in Mr. Hirst's report of egress that the interval in time between actual contact and his seeing the black drop was 1m. 45·41s., almost exactly the same time as ingress.*

I regret exceedingly that my eye was not at the finder during the precise moment of the formation of the drop, but my duties at the Janssen apparatus prevented me from staying there more than a few seconds at a time.

---

* The exposure of Janssen plate No. 9 was begun at.................................... 3h. 51m. 42·42s.
And it was finished at.. 3h. 52m. 31·39s.
Mr. Hirst took out the plate and looked in the finder and saw the black drop as described at .... 3h. 53m. 3s.
Mr. Vessey in the next tent observing with a first-class 4½-inch equatorial saw no drop, and contact did not take place until (Mr. Vessey's report) .... 3h. 54m. 48·41s.
hence it appears that at 3h. 51m. 42·92s. there was no visible black drop at 3h. 53m. 3s., the black drop was visible through the finder of the photo-heliograph; while it appears from Mr. Vessey's observations that actual contact as seen with a good telescope did not take place until 3h. 54m. 48·41s., so that the black drop was seen with the imperfect telescope 1m. 45·41s. before contact, and may have been visible a few seconds earlier.

Referring to what I saw through the finder I am convinced that my observations, short though they were, have not deceived me. I was thoroughly prepared, and on the look-out for the phenomenon at egress, and I have not the slightest doubt that any one, using similar optical instruments, would undoubtedly have observed what I did.

If we turn now to No. 5 Janssen plate (plate XIV) and seek a photograph of the drop, we find that photography, at least when aided by Mr. Dallmeyer's beautiful lenses, refuses to acknowledge any such phenomenon; on this plate there are sixty photographs without a sign of the drop, but all showing a distinct band of sunlight round the planet. It will be remembered that while this was going on in the photo-heliograph observatory, Mr. Vessey was in the next place observing the phenomena of ingress with a very fine $4\frac{1}{2}$-inch equatoral, by Schroeder. With this instrument a splendid view of the ingress was obtained, and he noted internal contact at 12h. 23m. 45·07s. No. 5 Janssen plate was begun at 12h. 25m. 35·47s., and Mr. Hirst saw the drop 5s. before this, or at 12h. 25m. 30s., or some time after ingress had taken place, and it appeared to him equal to rather more than one-third of the diameter of the planet. Now we know it was only 1m. 45s. after observed ingress, and the photographs prove that the planet was only 1-22 part of its diameter within the sun's limb.

Of course there is the possibility that the drop might have broken between the time when Mr. Hirst saw it through the finder of the photoheliograph and the time he began to turn the handle of the Janssen apparatus on the same instrument; and the time lost in this change could not possibly exceed 5 seconds, for he passed from one to the other as quickly as possible, and even if it did break, we have the facts clearly made out that the drop was seen 1m. 45s. after ingress, and that although it appeared nearly equal to the one-third of the diameter of the planet in length, yet it was certainly not more than 1-22nd of the diameter as shown by the photograph.

Mr. Allerding, chronometer maker, of Hunter-street, also saw the drop most distinctly, and watched it through the various phases till it broke. He was using at the time a very good $3\frac{1}{2}$-inch achromatic telescope, but to avoid sunlight and heat he had reduced the aperture to two inches, and with this small opening he obtained very satisfactory definition of the sun and planet. Unlike Mr. Hirst, who observed in the beautiful atmosphere of the mountains, Mr. Allerding observed from the back yard of his house in Hunter-street, which is surrounded by houses. In a report of his observations, which he has furnished to me, he says :— 'At the internal contact at ingress, I saw a drop which formed into a cone, and when this had nearly disappeared it seemed to stretch out to a fine thread (*see* plate XIII), to which Venus seemed to be attached. The thread appeared hard and definite, without any hazy margin, and I estimated its length at one-third the diameter of the planet. It then instantaneously disappeared at 12h. 24m. 44s., and Venus appeared already well detached from the sun's limb. Had I not waited for the disappearance of the fine line, I would have made inner contact at least 30 seconds sooner.' The mean of the time for internal contact given by 7 other observers in Sydney is 12h. 24m. 28s. Mr. Allerding made it 12h. 24m. 14s. Now, in this case we have no Janssen photographs to show how long the drop was, but my own observation taken in Sydney proves that the drop seen by Mr. Allerding was equal in length only to the space moved over by the planet in 45 seconds, that is, 1·7 seconds of arc from the sun's limb—that is, $2\frac{1}{2}$ times the length Mr. Stone estimated it to be.

In New South Wales, therefore, only those who were using telescopes of small aperture, $1\frac{1}{4}$ and 2-inch, and low power eye-pieces saw the black drop; and one, Mr. Hirst, was in a remarkably clear and steady atmosphere, and Mr. Allerding in a very unfavourable one, owing to the radiation of heat on a hot day from all the houses around him. So far then as this evidence goes the black drop does not seem to be due to the atmospheric conditions, but rather to the imperfections of telescopes of small aperture and low power.

There are, however, several observations recorded of a kindred phenomenon that I should like to place on record. At ingress I saw nothing of it, but at egress I saw it distinctly; and the cause is, I think, easily traced. But to take, first the observations at ingress.

Messrs. Belfield and Park, who were observing at Armidale with a $4\frac{1}{2}$-inch Cook telescope, that I examined, and know to be a good one, have sent me a valuable report and drawings of what they saw, and state that ' while Venus was advancing at ingress to about one-fourth her own diameter upon the sun, a faint tremulous shaking was seen between the limb of the sun and the planet (both bodies being very sharp in outline), which disappeared so gradually that it could not be said to have been obliterated at any particular instant. (*See* plates IX and XX.)

Mr. Bolding, P.M., Raymond Terrace, observed with a 3-inch telescope, and has forwarded to me a very complete report of the whole transit, and remarks:—' At the moment I expected the complete circle (*i.e.*, internal contact) came the apparent pause, instantly followed by a kind of indistinctness, which resolved itself into the form of a figure 8. The thing seemed to be holding up the planet, so to say, and appeared as represented in plate XXVIII, diagram 3. The line seemed blacker than the central spot; then the light came very distinctly between the planet and the line; then the indistinctness between the sun's limb and the line cleared up, and for a short time the line was clearly seen midway between the planet and the sun's limb. The sun was very hot at the time, and the definition bad.'

Mr. Russell: At ingress I saw nothing of the phenomenon, but at egress I did, and my report is as follows:—' At times there were moments of bad definition, evidently caused by the clouds then forming in the west. During one of these, at 3h. 53m. 54s., when Venus was less than two seconds of arc from the sun's limb, the limb of the planet nearest the sun's edge seemed to be in a state of vibration, as if portions of its blackness were jumping over to the margin of the sun with an appearance similar to sketch which represents one vibration only. This lasted only a few seconds—the vibrations being estimated at six or seven per second.

After this the limbs recovered their perfect definition, and were clearly and steadily separated by a fine line of light. Mr. Lenehan saw it, and says, ' The first apparent contact was at 3h. 54m. 22s., a little jumping. I afterwards saw a band or faint and narrow streak of light between the limbs of planet and sun.'

Messrs. Belfield and Park saw the same appearance at egress as at ingress. Mr. Bolding saw nothing of it at egress, which he attributed to the increased steadiness of the atmosphere.

I think there can be no doubt that this appearance was caused by temporary unsteadiness in the atmosphere, which, by producing rapid vibrations or apparent motions in the limbs under examination, caused them momentarily to overlap, and so cut off the sunlight and produce the black appearance, an effect which all who have been in the habit of observing with powerful telescopes will at once understand.

We come now to the last point that I propose to speak of. The information I have collected about it is in some respects very remarkable. I refer to the rings of light and especially the halo seen surrounding the planet Venus, a conspicuous phenomenon seen by nearly all the observers in New South Wales. That it was a very brilliant and beautiful object will be made abundantly evident by the accounts which follow.

And beginning, as before, with Eden. Mr. Scott, who was using a $7\frac{1}{4}$-inch equatorial, of very fine defining power, and of which the aperture was reduced to two inches, says:—'For some minutes before internal contact I could see clearly at ingress the whole of the planet's outline; in fact, it presented exactly such an appearance as might have been expected from a planet possessing an atmosphere.' Mr. MacDonnell says:—'At the time of apparent bisection, a shadowy nebulous ring seemed to envelop Venus (*see* plate XVIII; on the preceding side it was of lighter tint than the planet, but was decidedly perceptible, and appeared to be about one-quarter or one-fifth of the diameter of the planet in width. When ingress was about two-thirds completed, the whole outline of the planet was distinctly visible in the telescope, the shadowy envelope surrounding it very plainly.'

At Goulburn, Captain Onslow first saw the halo, or ring of light, at 12h. 17m. 5s. A bright light was seen at the lower point of intersection of the circles (*see* diagram plate I, figs. 1 and 2), and in a few seconds a similar one at the upper point, and at 12h. 19m. 5s. an apparent circle was formed by the planet.

Professor Liversidge says, when the planet was about one-third of its diameter from third contact:—'It then appeared spheroidal, and not as a disc merely (*see* plate XXII, figs. 1, 2, 3); it appeared illuminated on the under side in the direction of the sun's diameter, or on the side of the planet towards the sun's centre, and this illumination shaded off on each side of the planet, but at the portion nearest the sun's limb it appeared quite black and opaque. This globular appearance was retained until the planet had passed off the sun's limb to the extent of about one-sixth of its diameter.'

'After internal contact, the planet looked somewhat as if it were pushing that portion of the sun's limb before it, for the solar limb appeared to be raised up into two processes—one on each side (plate XXII, fig. 6). At the time I thought it might, perhaps, be due to an atmosphere surrounding Venus, or to an optical illusion; but since I have heard that other observers saw the illuminated edge of Venus beyond and outside the sun, I am inclined to think it was that which I saw. However, I did not see a circle, but merely two portions or cusps brightly illuminated, but not as bright as the sun.'

At Woodford, Mr. Vessey, who had the best atmospheric conditions and a first class telescope, saw so much of the shading on the planet and the halo that it would not be possible to reproduce all he says without extracting greater part of a long report. The shading on the planet was first seen at 12h. 7m., but not on the part off the sun; it appeared to extend inwards from the limb, resembling a gradually fading line of dots.

At 12h. 15m. 30s. the following limb of Venus was distinctly defined by a faint line of light or halo which was rather brighter on the northern side; 3 minutes later ring of light increasing in beauty, silvery, decidedly brighter on north side of middle, perhaps $\frac{1}{2}$ a second in thickness.

After complete ingress the definition was magnificent (*see* plate I, No. 3), and atmospheric ring or shading on, *i.e.*, within the disc of the planet similar to what I first saw at 12h. 7m., but broader, and gradually shading off towards the centre, to be traced all round, giving Venus an appearance of relief like an oblate spheroid, or rather a flattened dome standing away from the sun, the radius of the flattened part being about half that of the planet.

At egress Mr. Vessey again saw the ring of light directly contact was made, and steadily as the planet proceeded, at first like a small arch upon the sun's limb at 4h. 2m. 35s., the ring of light on planet appeared as a sharply defined line, and less than one second of arc in thickness, 6 minutes later, disc of Venus still continues undoubtedly a globe, and appearing slightly reddish or copper coloured (plate I, No. 4), like the moon in an eclipse, the sky adjoining intensely black, with the suspicion of a greenish tinge contrasting with the colour on the planet.

INTRODUCTION.

Mr. Du Faur, observing at Woodford with a 3-inch telescope, the eye-glass of which (after being smoked) was cracked by the sun, and therefore in a very unsatisfactory state for observations, still saw the whole of the planet when it was about two-thirds on the sun; and during the interval between internal contacts, had frequent opportunities of observing Venus with a $4\frac{1}{2}$ inch telescope after it had been carefully focussed on the sun spots, and saw Venus as sharply defined as it would be possible to represent it on paper, and perfectly black.

Mr. Russell says, 'At ingress I did not see the halo or ring of light round the planet (plate XXVII, fig. 3.) until 12h. 10m. 0s. It appeared only round that part of the planet not on the sun. It was very remarkable and beautiful, like a fringe of green light, through which the faintest tinge of red could be seen. It was densest near the planet, and seemed to shade off to nothing with a diameter estimated at one second of arc. It did not appear solid like the disc of the sun; but, like light in a dense vapour (*see* plate XXX), as ingress proceeded, the halo became more conspicuous; but I did not observe any want of uniformity in its diameter. At egress I saw nothing of the halo until 3h. 57m. 7s., nearly $2\frac{1}{2}$ minutes after internal contact. The halo was exactly similar to that seen at ingress, and the whole of the planet at this time appeared to me intensely black. The halo remained steadily visible for some time, but gradually faded, owing to increasing cloud causing a great increase in brightness of the atmosphere about the sun; and at 4h. 6m. 52s. I first observed that the surface of Venus was not black as it had been, but appeared as if covered with thin hazy clouds, somewhat thicker on the planet's northern hemisphere. At this time, the haze having much increased, I lost sight of the halo, and at 4h. 12m. changed the coloured glass for one of lighter tint, and at once saw the halo again, and for the first time noticed that it was irregular in diameter; it seemed considerably broader at the *north pole of the planet* and shaded off more rapidly towards B than C (*see* plate XXV, figs. 2, 3, 4, and 5), but I found it impossible to look at the sun steadily with this light glass, and again changed it for a darker one, when all the halo, except the part at the north pole, disappeared; this white patch continued visible against the sky (fig. 5.) until within one minute of last contact, and I feel confident I should have seen it some time after last contact but for the rapidly increasing atmospheric haze, which had also much increased on the planet, making it difficult to see where the haze on the planet ended and the sky haze began.

Mr. Lenehan says, 'At 4h. 16m. 21s. the planet appeared with the outer edge apparent, and I noticed a spot of light on the preceding side as at A. (*See* plate XVII, fig. 4.) It did not appear to me as anything more than a spot.'

Messrs. Belfield and Park saw the following limb of planet at ingress distinctly illuminated, and when the planet was wholly on the sun the body of the planet appeared intensely bluish black in centre, becoming gorgeous (*see* plates VIII, X, and XXI) deep blue towards the circumference; at egress the illumination of the planet's limb was again seen, but only on the north side.

Mr. Bolding only saw the halo at egress, and though visible all round that part of the planet off the sun was most marked on the north side. (*See* diagram, plate XXVIII.)

It will be seen that we have here three distinct phenomena. A broad ring of light outside the planet, a bright ring of light round that part of the planet projected on the sky, and band of light or shading round the inner edge of the planet, or over its surface. No spots, however, were seen on the planet, except the very remarkable part of the halo at the north pole.

The cause of the halo seen by Messrs. Lenehan and MacDonnell has not been satisfactorily made out, though it has been repeatedly seen during transits of Mercury. It seems exceedingly improbable that Venus has an atmosphere of such extent as would be required to produce such a halo or ring of light as that seen by both these observers. It appears, however, certain that it is one of those curious phenomena seen only by some observers under special conditions. The transit of Mercury in 1868 was watched very closely by a number of observers in England, who were seeking information that might be useful for the transit of Venus; and out of fourteen observers, including some of the best in England, only three make any mention of the diffused exterior halo. Mr. Stone thought it simply an effect of contrast.

Probably some of the light seen on the planet this time had a similar origin, no observer has, so far, reported seeing both. A part of it, however, must, I think, be attributed to haze in our own atmosphere, which, being very luminous owing to moisture then forming, would appear projected on the black planet, and the contrast would very likely give it a shaded appearance from the edge towards the centre. To me the blackness of the planet, both at ingress and egress, was very intense, until the haze in our atmosphere became thick and gave the surface of the planet a cloudy look, so that I could scarcely see where the planet ended and the sky began; and it may be that the same cause produced what Professor Liversidge and others saw; but at Woodford the air was too clear for such an explanation. The red tint seen by Professor Liversidge is explained by his having used a red glass shade.

The increase or thickening of the halo seen at the north pole of the planet, and which to several of the observers seemed to encroach on the planet, is a most interesting feature, especially if taken in conjunction with an apparent flattening of the planet, seen by Mr. Vessey at the opposite side.

The remaining ring of light or halo is the most interesting physical feature observed, though at first sight it would be attributed to an atmosphere similar to that of the earth. I think a little consideration will show that it cannot have such an origin. It is spoken of by all the observers as very brilliant, by some as white compared with the sun; and its actinic power was so great that, although its diameter was certainly less than one second of arc, and would only appear as a fine line in a Janssen photograph less than one five-hundredth part of an inch in diameter, it yet had power to effect the chemicals more than the sun itself in something less than the two hundred and fiftieth part of a second; in other words it was more powerful in affecting the silver salts on the photograph plates than direct sunlight, and we have a number of Janssen plate photographs in which it is shown by a deposit of silver thicker than that made by the sunlight.

This great brilliance, of course, explains why it was not seen about the planet while on the sun's disc. It was evidently not to be distinguished from the sunlight.

In the clear atmosphere at Woodford, it was seen as soon as the cusp parted at egress, and it will be exceedingly interesting next time Venus is lost in the sunlight to try if, under favourable conditions, his halo can be seen. Quite sure I am that, if the air had been clearer at egress, I should have seen the planet with the halo round it projected upon the sky, as it was I saw part of the halo until Venus was nearly all off the sun's disc, and one minute before last contact. Taking all these facts into consideration, I cannot see any cause sufficient to produce a halo or a ring of light such as that described, an atmosphere by refracting, would diffuse the light and by absorption would reduce it, so that the halo cannot be the result of an atmosphere.

Some of our Janssen plates give results which are obvious enough without measurement; one of these is the extreme sharpness of all the cusps. Of the sixteen plates having about sixty photos on each—the first plate shows a small notch in the sun's limb; Nos. 2 and 4, planet still further in; No. 5 is the one taken when the black drop was seen; No. 6 shows the planet wholly within at ingress 7, 8, and 9, the same at egress; No. $9\frac{1}{2}$ shows the planet on the sun's limb, with the halo; No. 10, planet partly off sun, with some pictures of the thickening of the ring of light about the pole of the planet; Nos. 11 to 16 plates, at egress; No. 17, was passing through when last contact was observed, and shows the faintest notch in the sun's limb till within a few seconds of observed last contact. The value of these plates is very great; photography is not biassed by preconceived theories of what it should see, and is therefore a witness upon questions of physical aspect whose evidence no one may gainsay.

On February 22nd, 1875, I left for England, taking full copies of all the reports and observer's drawings, together with ten Janssen plates, each having sixty pictures of a portion of the sun with Venus near the limb, and fifty-seven photographs of the sun four inches in diameter, some taken at Sydney and others at Woodford; these were all given to the Astronomer Royal for England, on July 14th, 1875, as the New South Wales contribution towards the observation of the Transit of Venus, and in January, 1876, others were sent, making up the numbers to 109 Sydney 4-inch plates and thirty-six Woodford plates, and twelve Janssen plates, some of these not having quite the full number of photographs on them.

The observations will be found very fully illustrated, and other plates have been added showing Sydney observatory, the temporary observatories used at other places, and also, the larger instruments, tents, &c. Unfortunately, Goulburn buildings were taken down before a photograph of them was secured and the only one of the large instrument is very imperfect but it is the best available.

H. C. RUSSELL,
Sydney Observatory,   Government Astronomer.
16 August, 1892.

MR. RUSSELL'S REPORT re TRANSIT OF VENUS.

Sydney Observatory, 9 December, 1874.

THE early morning was clear and fine, but from 5h. 30m. a.m. to 8 a.m. *The weather.* thick fog-like clouds covered the sky; they seemed to be very low down, and all melted away under the increasing heat, leaving the sky beautifully clear and promising for the work before us—a promise which was fully realized during the day.

All the observers were at work early, giving finishing touches to our *Preliminaries.* preparations, and giving and receiving clock signals to ensure accurate time. At Sydney this occupied a considerable time, for three stations— Eden, Goulburn, and Woodford—required the signals, and some were given to private observers. By 11 a.m., however, all this was over, and all the telescopes and photographic apparatus for use in Sydney were quite ready for the work, and the observers had time to look quietly over the preparations and see that all was ready.

The principal instrument was the new Equatorial by Dr. Schroeder, *Instruments.* of Hamburgh, which had been obtained for the purpose of observing this transit. It has a clear aperture of 11·4 inches and a focal length of 12 feet 6 inches, and was provided with a full battery of eye-pieces, and a polarizing apparatus for viewing the sun. The definition of this instrument is superb, with the new achromatic eye-pieces supplied by Dr. Schroeder, but owing to the great heat concentrated in the focus on that bright summer day, it was found necessary to reduce the aperture to five inches when observing and six inches when photographing. For the purpose of taking photographs it was fitted with a camera and enlarging lenses of such power that the sun's image measured four inches. The photo plates were placed simply at the end of the camera and held by a spring while the picture was taken, no dark box being necessary, because the camera end of the telescope passed into the dark room, which was simply a tent raised inside of the dome and connected with the telescope by means of a flexible sleeve, so that the telescope was free to move with the clock-work. A flashing shutter of the ordinary kind was used, and when the plate was in position, a very light spring was touched, set the shutter free and made a picture; which was immediately removed and developed, and as soon as it was washed, the shutter was lifted, and the camera ready for another. This precaution was necessary, because the shutter in its motion upwards, which was of course by hand, and comparatively slow, let a flash of sunlight into the dark room. Thus arranged it was found that three persons could and did work at the rate of one photo per minute, with the ordinary wet collodion process. One coated the plates and put them in the baths, of which four were used, fixed on a turn-table, so that by the time a plate had travelled round it was sensitized. The second worker took the plate out and put it in the camera, exposed it, and handed it to the third, who developed it and finished the picture; this duty devolved upon me, and I was thus able to see during the progress of the work that the plates were properly exposed, and that the driving-clock kept the sun's image on the middle of the plate. Attached to the flashing shutter apparatus was a contact-maker and two wires that led to the chronograph; every flash of the shutter was thus recorded, and against each record on the tape the number of the plate exposed was written, which thus furnished an exact record of the time of exposure, the plates being identified by numbers written on them with a diamond beforehand. The same chronograph was used for recording the times of the various phenomena observed.

Distortion.

Before I leave the instrument, it is necessary to say that the enlarging combination was of peculiar construction. It had been found that the ordinary enlarging lenses, and especially the one made for this telescope, gave to the limits of the field considerable distortion. To get over this difficulty experiments were tried, and it was found that two plano-convex lenses of equal focal length placed convex sides towards each other could be so adjusted that there was no distortion of the field; in fact a large white screen was carefully ruled into inch squares and placed four hundred feet from the telescope, when photographed, all the lines were straight. This combination was therefore adopted for the transit work, and all the Sydney photos were taken with it.

Enlarging lenses.

As a proof of its accuracy, the Scale of Inch squares was set up at a distance of 400 feet and a photograph of it taken, from which by direct printing the photo herewith was produced. (See plate XXXII.) An inspection of this shows that there was no distortion, or in other words, the field was quite flat.

The camera was so made that it could be put on or off the telescope in one minute. Before ingress the telescope was placed ready with a direct vision achromatic eye-piece magnifying 100 times, and the coloured glasses were a green one before the eye-piece and a dark blue or neutral shade near the eye; so protected, no inconvenience was felt from the sunlight. Other magnifying powers were tried, as was also the polarizing eye-piece; but the observations were made at ingress and egress with the 100 eye-piece described above. The telescope was clock-driven, so that my whole attention was given to the work of recording the times of contacts at ingress and egress, and to secure accuracy a chronograph on which the ticks were marked by the standard clock was used, and against each mark made by the observer, an explanatory note was made on the paper tape by the assistant in charge of the chronograph; and in order to guard against the total loss of the observation by failure in the electrical contacts, I had a chronometer near me and recorded the time to the nearest second by it. These notes served also as a check on the chronographic work.

The error and rate of the chronometer were found by comparison with the standard clock before and after the observations, and the error and rate of the standard clock were found by transit observations the nights before and following

As soon as I had observed the phenomena of ingress, the eye-piece was removed and the camera put in its place, and we commenced to take photographs of Venus in transit, and during 3 hours 10 minutes 190 pictures were taken.

The observations at ingress.

At 11h. 40m. a.m.* I began to observe, and adjusted the focus very carefully, so that spots and faculæ on the sun's surface could be seen distinctly, and then a close watch was kept on that part of the sun where the planet was expected. I was surprised I could see nothing of the planet at 11h. 54m., but kept my eye steadily upon the place, and at 11h. 55m. 23s. First contact saw the first sign of Venus certainly on the sun. I thought that the indent I saw was the planet a few seconds before, but could not be sure it was not one of the irregularities in the sun's limb due to atmospheric causes, until about 15° of the planet's circumference were in contact. Probably the time should be 10 seconds earlier. From this point until 5 minutes past 12 (noon) there was nothing to remark but the sharp and definite outline of the planet as it crept over the sun. At this time, 12h. 5m.,

* All the times given in this and other following reports are in Sydney mean time.

TRANSIT OF VENUS.

Planet all seen. the whole of the planet became visible (plate XXVII, fig. 1), that portion of it without the sun appearing on the bright sky near the sun's limb; as ingress went on, the planet became more and more distinct, and seemed to stand in relief on a back ground of greenish grey light (fig. 2). It was uniformly black, and I could not detect any haziness about its outline on the sun, except during moments of bad definition, which were temporary, and evidently due to changes in our own atmosphere. The outline of the sun and the cusps were also sharp, but I would not be certain that the halo afterwards noted was not then visible.

Halo. At 12h. 16m. I first observed the halo (plate XXVII, fig. 3); the planet had been getting gradually more conspicuous both on and off the sun's limb, and my attention had been principally directed to the cusps, to detect, if possible, any phenomena like the formation of the D-shape, but nothing was seen, and when taking a general look at the time noted, I first observed the halo round that part of the planet not on the sun. It was very remarkable and beautiful, like a fringe of green light, through which the faintest tinge of red could be seen; it was brightest near the planet, and seemed to shade off to nothing, with a diameter estimated at one second of arc. It did not appear solid like the disc of the sun, but like light in a dense vapour. As ingress proceeded the halo became more conspicuous, but I did not observe any want of uniformity in its diameter, and it at all times terminated at the sun's limb, there being no sign of the halo on the part of the planet on the sun.

Plate XXX is an attempt to show what was seen, but the halo is far too broad, and I was unable to put on paper what I saw.

Bad definition At 12h. 20m. indications of distortion or bad definition of the limbs in contact appeared, like a mass of black wool laid over the place (plate XXVI), rendering it impossible to see distinctly, and making the cusps very hazy. I thought the drop was going to form, and watched very closely for it and for apparent contact; but I found it extremely difficult to make up my mind about the latter, and saw nothing of the former.

Apparent contact. 12h. 20m. 51s. was noted a very unsatisfactory apparent contact. The cusps after this appeared to clear up or improve in definition, and as they approached each other the sharpness was very remarkable, but the motion so gradual that I could not determine to a fraction of a second when they actually completed the line of light; but the instant I *Internal contact.* was sure I made the record on the chronograph, for "internal contact";

the time was 12h. 23m. 59s., and keeping my eye steadily upon it, saw it had in 15 seconds become a distinct and unmistakeable band of sunlight. The planet and its neighbourhood were then examined with great care, and presented a perfectly black unmarked surface, with a hard and distinct outline, and near it nothing but the uniform light of the sun could be seen.

Observation was then (12h. 30m.) given up, and the camera put Photography. upon the telescope, and in the course of 3 hours and 10 minutes 190 photos were taken. Four of these are useless from accidental causes, but on the remainder is a clear and beautiful record of the planet's progress across the sun.

Egress.

At 3h. 40m. p.m. we gave up taking photographs and prepared for observations. The same eye-piece and coloured glasses were used. The planet now had a perfectly sharp and clear outline, that of the sun being also very good. I could not after careful scrutiny see anything remarkable on the margin of or near the planet, and the limbs continued to approach each other beautifully defined (plate XXIV fig 1). Great care was exercised to keep a steady watch without straining the eyes. At times there were moments of bad definition, evidently caused by the clouds then forming in the west. During one of these, at 3h. 53m. 53·59s. the limb of Jumps. the planet nearest the sun's limb seemed to be in a state of vibration, as if portions of its blackness were jumping over to the margin of the sun with an appearance similar to fig. 2, plate XXIV, which represents as nearly as can be estimated the space over which the phenomenon was seen as compared with the diameter of the planet. This lasted only a few seconds, the vibrations being estimated at 6 or 7 per second; after this the limbs recovered their perfect definition, fig. 3, plate XXIV, and were clearly and steadily separated by a line of light, which at 3h. 54m. 26·30s. could not have Fine line of light. been more than half a second of arc in thickness; and then the same marvellous definition continuing just when it was wanted, the line gradually contracted to a scarcely visible thread (fig. 4), plate XXIV, and the limbs Internal contact. made contact at 3h. 54m. 39.66s. There was no sudden break—nothing but the perfectly gradual motion of the one disc over the other, both being beautifully defined, and the observer saw one limb overtake the other. I have no doubt that had the bad definition continued, the moment when the jumps were seen would have been taken as the formation of the drop.

A curious phenomenon then presented itself similar to that remarked at ingress,—the two limbs at the point of contact seemed to get confused or badly defined, whether from atmospheric causes near us or from some peculiarity about Venus I am unable to say, but it seemed to distort the planet in a most remarkable way (plate XXV. fig. 1, and plate XXVI). And now, as at ingress, I found it very difficult to determine apparent contact, or what under the strange effect, seemed like the phenomenon known by that name.

Apparent contact.   At 3h. 55m. 45s. it appeared to me that if Venus could be made round it would be just in contact with the sun's limb. I took particular note of the circumstance, in the hope that it would throw some light upon the so-called black drop.

Halo.   I did not at this time observe any halo or anything to indicate the exact position of the planet's margin, and it was not until 3h. 57m. 7s. (plate XXV, fig. 2 at B) that I saw the outline off the sun with the halo round it exactly similar to that seen at ingress, and now as then only on that portion of the planet which was off the sun. The whole of the planet then appeared to me intensely black (as at A, plate XXV, fig. 2); at 3h. 59m. 12s. the planet was all visible and the cusps very sharp; $2\frac{1}{2}$ minutes later the halo was only just visible; at 4h. 3m. definition bad, and the bright haze about the sun noted to be on the increase. Definition improved again,

Clouds on planet.   and at 4h. 6m. 52s. I first observed that the surface of Venus was not uniformly black as it had been, but appeared as if covered with thin hazy clouds, thicker somewhat on the planet's northern hemisphere, but nowhere sufficiently dense to prevent me from seeing the dark planet at the back, at least such was the impression formed at the time; at 4h. 12m. the definition again became bad. Since the last note about the halo it had almost disappeared, and I changed the dark blue shade for one of a lighter tint; with this saw the halo distinctly all round the part off the sun, but could not look steadily at the sun, the light being too strong; the halo was for the first time seen irregular—in diameter it seemed considerably broader at the north pole of the planet as shown (fig. 3, plate XXV, at A), and shaded off more rapidly towards B than C. At 4h. 15m. 20s. (plate XXV, fig. 4), the dark glass having been replaced, this northern band of light was all that was visible of the halo, and the haze on the planet was greater; at 4h. 10m. 2s. the cusps were not sharp but rounded off; the appearance was coincident with another turn of bad definition, and the atmosphere became now so much disturbed that there was little hope of a satisfactory observation of last contact. The white patch on the planet had however continued visible since it was first seen, though at times it became very faint; by its aid I was able to make out the outline of the planet, now a very difficult matter, for the haze on the planet had become almost as bright as the sky, and must I think have been due to the moisture gathering in our own atmosphere.

At 4h. 22m. 12s. the white patch was distinctly visible; definition good again; faintly seen at 4h. 23m. 22s. (plate XXV, fig. 5), and at 4h. 24m. 27s. the last sign of the planet on the sun's disc was seen as the faintest possible mark, which then disappeared; definition being for the time very good, the observation was quite satisfactory. After this I tried to see the white spot on the planet but failed, as the haze in the sky was rapidly increasing. *White still visible. Last contact.*

At this time the whole sky was very hazy, and long tapering clouds were coming up from S.W., and I have no doubt the moments of bad definition were caused by the passage of the points of these clouds across the line of sight.

H. C. RUSSELL,
Government Astronomer.

Sydney Observatory,
9 December, 1874.

TRANSIT OF VENUS.

Copy of Memoranda on Cards used during observations.

INGRESS :—

NOON. Planet perfect black clear outline (*i.e.* on sun).

h. m. s.	
12 4 0	Planet very clear and sharp, nearly bisected.
12 5 0	Can see the whole of the planet, but not clearly.
12 10 0	Whole of planet plainly visible on a back ground of faint grey light.
12 15 0	Still visible outside sun ; planet's edge seems perfectly sharp ; just a shadow of uncertainty about edge, think it is atmosphere.
12 16 0	All visible, at times beautifully ; no small body near Venus ; cusps perfectly sharp. (Halo was marked on chronograph at this time.)

EGRESS :—

3 54 30·00	First contact ; sharp contact of limbs ; 40s. later distortion about the point of contact.
	Venus looked in apparent contact with sun's limb.
3 57 7	Venus' margin off sun clearly seen with halo round it similar to that seen at ingress.
3 59 12	All planet visible ; parts on and off the sun, latter with halo round it.
4 0 22	All visible, but halo not distinct : bright margin off sun not so marked.
4 1 42	All visible, but only faintest halo.
4 2 52	Planet's margin all visible off sun ; not well defined ; haze about sun seems to be increasing.
4 5 12	All visible ; sun's edge very steady in glimpses.
4 6 52	Venus does not look uniformly black on north side—it looks as if planet was cloudy.
4 8 52	All Venus visible.
4 10 52	All visible.
4 11 52	Bad definition, but all planet visible.
4 12 52	(Changed sun-shade to lighter one.) All visible ; halo distinct, specially on the north side.
4 15 12	(Dark glass again.) Halo on north side only—it is very remarkable.
4 16 2	Cusps not sharp—seem rounded off.
4 17 42	White place still faintly visible ; white haze increasing.
4 18 52	Definition getting very bad—can still glimpse white spot by its aid follow the outline of the planet.
4 20 32	Still visible, but very faint.
4 22 12	Can still glimpse it.
4 23 22	Doubtful (glimpses).
4 24 27	Last contact ends, definition good.
4 25 52	Certain I can see nothing outside sun.

H. C. RUSSELL,
Government Astronomer.

TRANSIT OF VENUS.   9

REPORT FROM MR. LENEHAN.

Government Observatory, Sydney, New South Wales,
H. C. Russell, Esq., 12 December, 1874.
Government Astronomer.

Sir,
I have the honor to report result of my observations of the transit of Venus on the 9th instant.

The instrument used by me was an old telescope by Troughton & Sims of London, with a focal length of 6 feet 9 inches—an 80-power eye-piece, and neutral shade glasses of different densities—an aperture of $4\frac{3}{4}$ inches, stopped down to 4 inches by a cap over object-glass; fitted on equatorial stand and under cover of a temporary observing dome. *Instrument.*

At about 11h. 45m. I took my station at the eye-piece and attentively watched for the first appearance of the planet, having everything in my favour for a good result—definition of telescope, weather and clearness of atmosphere, all that could be desired, with the scattered spots upon the sun's disc showing sharp and clear. *Work.*

The first indent on the sun's edge by the planet was observed at 11h. 55m. 36·34s., although I fancy from the contact then formed I could have seen it 10 or 15 seconds earlier had I known the exact spot of entrance. *First contact.*

The planet had crept on the sun's disc about one-fifth its diameter at 11h. 58m. 21·34s., definition clear and sharp.

At 12h. 1m. 31·34s. about one-fourth on, and at 12h. 5m. fancied I saw the outer edge of planet, but was not perfectly clear on that point.

The sun's limb had bisected the disc of Venus at 12h. 7m. 8s., still clear and well defined; at 12h. 12m. 30s. it looked as if the edge of planet was losing its curvature, but later I found I was mistaken, as the cusps reformed, giving the true curve to the planet. *Bisected.*

About three-fourths of the planet on the sun at 12h. 15m. 54s., edge of planet off the sun discernible, but no drop formation. *All planet visible.*

At time of ingress there was an indistinct shading between the supposed edge of planet and the sun, which for some ten or fifteen seconds before the time I quote later, kept me in a state of uncertainty as to true

TRANSIT OF VENUS.

Shading. time of actual ingress. The shading did not break abruptly, but seemed to melt away in such a manner as to leave a doubt in my mind of the exact **Band of light Ingress.** time the planet passed the edge of the sun, but I distinctly saw a clear band of light at 12h. 24m. 48·31s. Then I left the telescope for the photographic room, preparing the plates for exposure in large equatorial until about ten minutes before first contact of egress, when I returned to my telescope on the north side of the Observatory to observe the egress.

The planet stood out on the sun's disc with a clear and sharp outline, with a luminous appearance or halo outside it about one-third diameter of the planet (plate XVII), of a greenish yellow, with outer edge an orange shade, as if the planet had an atmosphere, or perhaps caused intense brilliancy on that part of the sun's face by rays being reflected. The atmosphere was clear, and the following observations were made under very favourable circumstances.

There was a similar indistinctness between the time I mention as "apparent contact" and "complete contact" as noted in the former portion of this report at time of ingress.

Egress. First contact. The first apparent contact was at 3h. 54m. 21·61s., a little jumping *(plate XVII, fig. 2), afterwards saw a band or faint and narrow streak of light between limb of planet and sun which clearly showed the time given was in error (plate XVII, fig. 3). I then waited until I was *absolutely certain* contact was completed, which was at 3h. 54m. 46·61s., I feel confident that this time is from seven to ten seconds after true time; thinking the light might again show between made me wait till I was quite certain before I noted the second time.

Cusps indistinct. At 3h. 56m. 26·61s. the points of the sun cusps appeared blunted as if from vibration, and at 3h. 58m. 6·61s. this formation or appearance had somewhat increased (this formation was like capillary attraction between the planet's limb and the dark sky). The planet was about one-fourth its diameter off at 4h. 0m. 11·61s.; at two minutes later the planet's edge off **Halo.** sun was apparent. Half off at 4h. 9m. 1·61s., the halo still showing around that portion of the planet on sun's disc; about three-fourths off, at **Spot of light. Last contact.** 4h.16m. 21·61s. with the outer edge apparent, and noticed spot of light on preceding limb of planet, as shown at A (plate XVII, fig. 4); did not appear to me as anything more than a spot.

The last and final contact at 4h. 23m. 40·61s., the edge of planet being then lost in edge of sun.

---

* In plate XVII the diagrams should have been numbered 1, 2, 3, 4 from right to left.

These observations were made through a dark neutral shade eye-piece, and were not at all trying to my sight.

Shortly after the finish of observations the atmosphere seemed to become very smoky and thick, but altogether Sydney was favoured with very fine weather.

I have the honor, &c.,
HENRY A. LENEHAN,
Computer.

REPORT FROM MR. SAVAGE.

H. C. Russell, Esq., B.A.,   Sydney Observatory,
     Government Astronomer,—   21 December, 1874.
Sir,
     I hereby give you a short report of what came under my notice during the transit of Venus on December 9th, 1874.

The telescope allotted to me was a 10-inch reflector by Browning, of London, with an unsilvered speculum, and focal length of 6 feet 2 inches, with an eye-piece magnifying 200 diameters; a dark glass slide was used to protect the eye from the intense heat of the sun, the day being very hot. Shortly before the time for first contact I was at the telescope, but the instrument being a reflector and strange to me, caused me to lose the observation before I could get the telescope to bear on that portion of the sun's limb where first contact took place. *Instrument.*

The next point for observing the planet was when it was bisected, the time of which I noted as 12h. 6m. 33·93s. and the apparent internal contact at 12h. 22m. 1·93s., the definition at this point being very bad between the limbs of the sun and planet. *Bisected.*

As the planet advanced on the sun's disc a little way a shading connected the planet with the sun's edge, but that portion of it nearest to the planet showed indications of fading away gradually, until at length it disappeared altogether without any sudden break whatever, and at 12h. 23m. 43·93s. a streak of light became visible between the planet and sun's limb. *Shading of cusps.*

Egress.

Polar spot.

For the same reason as above stated I did not observe the first contact at egress; the planet when I got it fairly in the field was about ¼ of its diameter off the sun. The bisection of the planet took place by my observation about 4h. 5m. 49·68s. I did not see any outline of the outer edge of the planet during egress, but I did see a momentary flash of light in the position shown in plate XI, but I did not record the time. The last external contact of the planet with the sun's limb I recorded as being 4h. 19m. 32·68s.

During the ingress and egress the definition of both the sun and planet was very fine. Through my not being able to get the first contact at ingress and egress the clock motion was dispensed with, and I moved the telescope gently by hand as it required. The interval between ingress and egress I was engaged in equatorial tower, keeping time records of each photograph taken.

EDWIN GEORGE SAVAGE,
Meteorological Assistant,
Government Observatory,
Sydney.

WOODFORD REPORTS.

REPORT FROM THE SURVEYOR-GENERAL *re* TRANSIT OF VENUS OBSERVATIONS AT WOODFORD, ADDRESSED TO THE GOVERNMENT ASTRONOMER.

Surveyor General's Office, Sydney, 21 December, 1874.

Dear Sir,

Position of Woodford.

I have to report the favourable observation of the transit of Venus, at the temporary Observatory at Woodford, on the Western Railway, 54 miles west of Sydney, 2,200 feet above the sea, latitude 33° 43′ 58·7″, long. 10h. 1m. 56·20s. (for details see Appendix 5), and to append the reports of Messrs. Vessey, Hirst, and Dufaur.

The duties of each member of the party of observation were taken *Division of duty.* up as nearly as possible in accordance with your wishes; and I much regret the breaking of the glasses dark of the 3-inch Cooke telescope, by which we are almost deprived of the value of Mr. Dufaur's observations.

Leaving Sydney on Monday, the 23rd ultimo, I was employed until *Preparation.* Thursday afternoon in marking out the position to be occupied by each observatory, unpacking and attending to the instruments, which were thoroughly wet by rain which fell at the time the packages were put down by the train, and which continued the three following days; little progress could therefore be made in the erection of the piers, the earth being saturated with wet, so I returned to Sydney, and revisited the camp on the 30th ultimo, accompanied by Mr. Hirst, and was followed the next day by Messrs. Vessey, Bischoff, and Dufaur. The day after we were fully employed in setting up instruments, consisting of the photoheliograph by Dallmeyer, the siderial clock by Cooke & Sons, a chronograph, the 4½-inch Schröder telescope, kindly lent by A. Fairfax, Esq., a 3-inch telescope by Cooke, and a portable transit instrument—the two latter being supplied by the Survey Department.

The place of observation was about 200 yards westerly from Mr. *Situation.* Fairfax's house (Woodford), and the same distance northerly from the Western Railway and Telegraph line, and nearly upon the summit of the main dividing ground between the waters of the river Cox on the south and the river Grose on the north. So narrow was the ridge in the direction of the meridian, that we could not get a tree or any object as a referring mark without crossing the valleys, which would have made the mark most inconvenient, and it therefore became necessary to erect a theodolite upon the rock, within 70 yards of the transit instrument, to be used as a referring mark, by observing one of its wires with the transit instrument.

Until the 5th the weather was most unfavourable for observations of *Weather.* any kind, but opportunity was taken of every fine opening to adjust the photoheliograph, which, being of peculiar construction, was not so easily *Instruments.* adjusted as an ordinary equatorial. During this time ample opportunity was afforded of testing the value of the improvements made by yourself upon this instrument since it was imported into this Colony, especially the substitution of an eye-piece for adjusting it instead of the ground glass-shade. With this we were able to get a fair image of a star in the same focus as the image of the wires, so that the instrument with camera attached became almost as easy to manage as an ordinary equatorial, and the adjustments of both the stand and reference lines were made perfectly satisfactory.

The arrangement also by which the object-glass of the photoheliograph was made to pass through a sleeve of non-actinic calico attached to the shutter opening in the roof converted the observatory into a photographic room, and gave us every facility for taking the pictures quickly; because it was not necessary to use a dark slide to transfer the plates to the camera; they were taken out of the bath and put at once into the frame which had been fixed in the camera to receive them, and ensure an accurate focal position for all the plates. We found the lever and spring attached to the flashing shutter much more convenient to work than the cumbrous arrangement for working the shutter with which the instrument arrived here.

Telegraphic communication. On the 6th instant we were, through the kind co-operation of the Superintendent of Telegraphs, in electric communication with Sydney Observatory, a matter of considerable moment, as we were, through the *Difficulties owing to bad weather.* unfavourable state of the weather, all behind with our transit observations. and telegraphic communication enabled us to compare our time with the clock in the Sydney Observatory. The telegraph instrument was set up in a tent close to both the clock and transit rooms. After transit observations on two successive evenings for time and longitude, I found, on the 7th instant, when preparing for observing sun's meridian passage, that the sixth wire had broken and become entangled with two others. I restored them to position and supplied the missing one, but only to be again disappointed, wire No. 4 breaking before we had any further observation, so I determined on sending the instrument to Sydney for new wires, and make the observations for latitude and longitude at convenience.

Weather on 9th. The morning of the 9th was ushered in with a dry, hot, westerly wind, which increased to almost a gale at 10 o'clock. Shortly after 10 we received your latest time signals; and after wishing success, and asking further signals as soon after the transit as possible, we each took our stations, from which time I have personally no knowledge of any occurrences beyond the interior of the photoheliograph and chronograph room, *Temperature.* the temperature of which was noted half-hourly. The result thereof appears as Appendix No. 3 to this report. My greatest anxiety was for the *Anxiety for even temperature for clock.* maintenance of as even a temperature for the clock as possible, and but for the continual revolution of the fan ventilator and the constant application of wet blankets, we should have found it difficult even to exist in the closed room. The temperature, however remained as steady as could be expected under the circumstances. For the observation of time of ingress and egress and the phenomena observed with the $4\frac{9}{10}$-inch telescope I refer you

to Mr. Surveyor Vessey's report; and from his skill and practice in delinea- _{Mr. Vessey.}
tion of form, and its relationships, I have great confidence in his observations, especially of the egress after he had time to realize the entire absence of expected phenomena.

I have next to invite your attention to the somewhat conflicting _{Mr. Hirst.}
report of Mr. Hirst, given as Appendix No. 2, who distinctly saw a "drop" _{Black drop.}
whilst observing for position through the finder of the heliograph, and in reply to my inquiry as to its appearance, said "more like a stalk." I regret now that I did not leave my work to witness the appearance myself, but thinking that others would see it, I did not wish a break in recording. I cannot account for the appearance reported by him, unless it arose from some optical delusion, resulting from the boiling appearance of the limbs of both the sun and planet when seen through a telescope of small aperture with an inferior lens. My experience of Mr. Hirst fully realized your expectations of him as a gentleman of untiring zeal.

I have next to invite your attention to a short report from Mr. _{Mr. Fairfax.}
Fairfax, to whose kindness and ready assistance we owe so much. Mr. Fairfax, though not actually taking part in our work, was yet present giving every assistance, and when the telescope was not otherwise in use he took advantage of the interval to get a look at Venus. His report, though short, is important, bearing as it does directly upon the character of the ring of light seen round Venus. From his keen sight and long _{Halo.}
experience as an amateur observer I attach much weight to his report. [Appendix 7.]

Before leaving the subject, I should like to say that, even if we have not succeeded quite as well as we had hoped to do under more favourable circumstances, still I shall always look back upon our expedition with great pleasure. We agreed together well throughout, and can all bear testimony to the kindness of Mr. Fairfax in placing his house and servants at our disposal, and last, not least, the use of his excellent telescope.

During some of the intervals of weather favourable for observation _{Clearness of atmosphere.}
I saw some of the groups of stars in Argo in a degree of perfection rarely to be hoped for in Sydney, fully bearing out the high estimate you place on the Blue Mountains as a point of great advantage in observing celestial phenomena. I remain, &c.,

P. F. ADAMS,
Surveyor-General,
N. S. Wales.

APPENDIX No. 1.

REPORT of Observations made by L. A. Vessey during Transit of Venus, 9th December, 1874, at Woodford, Blue Mountains, N.S.W., transmitted to the Surveyor General.

Instrument. INSTRUMENT, a 4¾-inch refractor by Schröder, stopped to 4 inches aperture, and equatorially mounted in a canvas observatory with revolving dome.

Eye-piece diagonal, with one greenish neutral glass, and power 96.

Chronometer. Sidereal chronometer by Frodsham, repeatedly compared with sidereal clock; times reduced to Sydney mean time. For convenience of observation and for the purpose of verifying apparent irregularities in the shape of Venus the diagonal eye-piece was used in different directions during the transit. At commencement of ingress it was pointed north, after ingress it was used for the most part pointing west, this being the position in which it was screwed home.

No shape.

Unsatisfactory contacts. The time observations at the internal contacts are very unsatisfactory. At ingress a gust of wind causing the telescope to vibrate fully half the diameter of Venus prevented a clear view of the phenomena, and the exhaustion caused by heavy night work in giving assistance with the adjustments of the photoheliograph, brought about a certain amount of nervousness in the observer at the critical moments, which was not lessened by the totally unexpected nature of the phenomena that occurred.

Clear idea of the phenomena. At egress the time observations were better, but the imperfect observation of ingress did not give the observer a clear idea of the phenomena he might expect at egress, and he was somewhat flurried by their gradual succession and the consequent impossibility of determining the precise moments of their occurrence.

First contact. Indentation of planet first seen on disc of sun, but unfortunately no attempt was made at the time to estimate the length of the notch. Time, 11h. 55m. 14·06s.

Boiling. There was a good deal of boiling on the limb of the sun, but not quite so much on Venus.

The planet appeared slightly flattened on that portion of the limb nearest the sun's centre, and with a slight bulge near northern termination of limb. (Plate I, fig. 1.)

The eye-piece was now changed for one of power 195, with two dark glasses, but the definition not proving so good the 96 was resumed, and the diagonal pointed west,—*i.e.*, the observer sat facing west.

Half-ingress. Estimated time of half-ingress, 12h. 7m. 0·00s. Previous to this a slight halo 3" or 4" within the disc of the planet was seen extending about 60° on each side from the limb of the sun, and resembling a gradually fading line of dots about 2" wide. (Plate I, fig. 2.)

There was no similar appearance on the limb of the sun, and the phenomenon remained constant, although the position of observer's head and the direction of the diagonal were repeatedly changed. The definition at this time was very good, the boiling practically nil, and the sun's limb wonderfully sharp, with the very slightest glare, which was evidently merely ocular.

TRANSIT OF VENUS. 17

The following limb of Venus distinctly defined by a faint line of light, which was *Ring of light.* rather brighter on northern side. (Plate I.) (Fig. 3 at 12h. 15m. 30·00s.*) This line of light very distinct. Making Venus a complete circle.

Definition good, very little boiling.

Ring of light increasing in beauty, silvery; decidedly brighter on north side of middle *Inequality in ring.* sketch, perhaps ¼" in thickness. (Fig. 4, Plate I, 12h. 19m. 30s.)

Apparent internal contact estimated too soon,—12h. 23m. 27·12s.   *Apparent internal contact.*

Time of ingress taken,—12h. 23m. 47·07s.

Ingress certainly complete,—about 12h. 24m. 0s.   *Real internal contact.*

The original note here is—"the telescope was much shaken by wind at 12h. 23m. and afterwards, and the ring of light around Venus cut through what would otherwise, I think, have been a drop, lasting 40 seconds, if not a whole minute." That this was a mistaken impression, and the natural result of the motion of the telescope, the notes at egress will clearly show. (Plate II.)

Definition magnificent at 12h. 30m. 30·4s.; atmospheric ring on disc of planet, corresponding to the halo noticed before (see fig. 2, plate I), but broader and gradually shading off *Globular appearance* towards the centre, to be traced all round, giving Venus an appearance of relief like an oblate *of planet.* spheroid, or rather a flattened dome standing away from the sun, the radius of the flattened part being about half that of the planet. (Plate VII, fig. 1.)

The observer then entered the photoheliograph room to assist in taking the large photographs, and had only an occasional hurried glance through the telescope till near the time of egress.

Definition very good, 3h. 46m.

The note made at ingress shows that the observer was, by his imperfect view of that *Egress.* phenomenon, fully prepared to see a drop of perhaps 2" in length divided by the ring of light around the planet, but the appearances that did present themselves at egress were very different, and to the observer certainly gradual, without the slightest suspicion of a sudden change or break in the light between the limbs of the sun and planet; the sun's light diminished till nothing but the silvery ring of light around the planet was left, and this ring was carried out on the sky by the slow motion of the planet.

About this time, 3h. 54m. 25·47s. (see plate IV), the light between the limbs of the *Apparent internal con-* sun and the planet became so thin that in observer's estimation all direct sunlight was cut *tact at egress.* off and nothing but the ring of light around the planet remained; or in other words, the narrowing of the line of sunlight between the sky and planet then seemed to cease.

At 3h. 54m. 37·50s. the cusps of the sun's limb were distinctly separated but connected by the fine ring of light around the planet. This ring was rather whiter than the sunlight, and the cusps at their extremities appeared very slightly thicker than the ring.

This thickening was only just perceptible, and might possibly have been caused by fatigue of the eye, requiring relief by a slight change of focus, or even by an unconscious mental influence, leading the observer to see what he was expecting and wishing to see. (Plate V.)

At 3h. 54m. 48·41s. the circles of the sun and planet tangential, and the ring of light *Internal con-* about its own thickness outside the limb of the sun. (Plate VI.) (See next page, halo *tact.* estimate as 1" in diameter.)

---

* The time on Fig. 3, Plate 1, is in error, it should be 0h. 15m. 30s.

C

Apparent internal contact decidedly past.
The ring now distinctly showing as an arch outside the sun's limb, 3h. 55m. 3·37s.

Venus not circular. Ring.
At 4h. 2m. 30s., Venus not quite circular, slight flattening on the eastern side of the planet as shown in sketch, ring of light brighter near limb of sun at eastern side, or at N.N.E. part of planet appearing as a sharply defined line without haze or irradiation, and less than 1″ in thickness. (Plate VII, fig. 2.)

Planet copper-coloured.
At 4h. 9m. 0s., disc of Venus still continuous, undoubtedly a globe, and appearing slightly reddish or copper-coloured like the moon in an eclipse; the sky adjoining intensely black, with the suspicion of a greenish tinge contrasting with the colour on the planet. (Plate VII, fig. 3.)

Repeated comparisons were made between planet and sky to verify this.

Half egress.
At 4h. 7m. 44·37s., probably estimated too early and a difficult observation to make, as a temporary shift of wind brought the heated air from the roof of the photoheliograph building across the sun, causing a sharp jagged boil.

At 4h. 17m. 0s., Venus not quite circular, the curve of planet's limb slightly flattened on eastern side, with a slight bulge on western side. (Plate VII, fig. 4.)

Final contact.
At 4h. 24m. 3·40s., last contact at egress. This observation appeared to be correct to a small fraction of a second. The indentation on sun's limb gradually contracted in width till about 12h. 23m. 51s., it then seemed to contract longitudinally till it became a small notch like a boiling indentation. This was seen steadily diminishing till it suddenly flashed out and the limb of the sun became perfect at 4h. 24m. 3·40s.

Definition very good, and boiling of sun's edge scarcely perceptible.

L. ABINGTON VESSEY,
Licensed Surveyor, New South Wales.

APPENDIX to Mr. Vessey's Report, with complete drawings of phenomena observed, transmitted to the Government Astronomer, by L. A. Vessey, Licensed Surveyor.

Scale of diagrams.
THE thirteen diagrams which accompany this Appendix have been drawn, the smaller ones on a scale of 40″ = 1 inch, and the larger 40″ = 3 inches, and the phases of the planet and points of the compass have been laid down with all needful accuracy. The drawings have been executed after much consideration, and represent to the best of my ability the appearances presented in the telescope; the smaller ones should be viewed at a distance of 5 feet, the larger at about 10 or 12 feet.

Drop phraseology.
In my report I unconsciously adopted the phraseology I had become accustomed to when using the artificial transit; the drawings will prevent any ambiguity that might arise from this cause at the critical times of internal contacts, and will make the report more precise throughout.

Through pressure of professional duties and the large call upon my leisure, made by the necessary preparations, I did not read up accounts of past transits carefully, except as referring to the "black drop," but gave my entire attention to acquiring accuracy in taking time and the estimation of tangential contact with the artificial transit. I did not expect there would be any physical appearances to record except at the critical moments, and I determined then to make sure of my times in the first place, and neglect (if need be) anything else that promised to interfere with that object.

Free from bias.
I therefore went to the telescope perfectly free from mental bias, and with the exception of the slight eccentricity shown in plate I, fig. 1, watched the planet creep, black and uninteresting as the sky itself, upon the sun's limb.

TRANSIT OF VENUS.

At the time when my attention was specially drawn to the cusps of the planet in the endeavour to estimate the time of half ingress I saw the halo. (Plate I, fig. 2.) The idea of an atmosphere instantly presented itself, but was succeeded as quickly by the thought that the halo was only an optical illusion, and I waited impatiently till the half ingress was past and I had leisure to test this. I then revolved the eye-piece 90°, changed the position of my head, and carefully compared the limbs of the planet with the adjoining limbs of the sun, and was at length satisfied that the halo was a reality and not telescopic; the sun's limb was sharply defined with only a suspicion of a glare as indicated in the drawing, in no way resembling the halo on the planet. The phenomenon remained constant, and I watched it steadily for some minutes, in fact to the best of my recollection until I caught sight of the ring of light shown in fig. 3. The beautiful pearly yet sparkling light of this ring would alone have been sufficient to fix my attention, even if its increasing brilliancy had not made plain that the observation of the expected black drop would be rendered very difficult when complicated by the presence of such a bright edge to the planet. Figure 3, plate I shows the northern segment of the ring brighter than the rest; fig. 4, plate I the north part decidedly brighter and corresponding exactly to the bright part in plate VII, fig. 2. *Halo on planet. Halo on limb of planet.*

The halo was not looked for, and was not seen till the time (plate I, diagram 3) when it appeared under a somewhat different form and was continuous round the planet; at egress it was not specially noticed. My note on (fig. 3, plate VII) "globular" seems to point to something of the kind in addition to the copper color, but not being certain of this I have not shown it in the drawing. *Polar part of ring.*

Plate II illustrates the phase of ingress which I endeavoured to catch as the "apparent" or tangential contact of the limbs of planet and sun (in the drawing they slightly overlap to give the desired effect). For a minute succeeding this a most untimely gust of wind kept the telescope vibrating, and plate III is only given as a representation of what I thought I saw during this time. There was much dimness and uncertainty about the cusps, whether from the motion of the telescope or otherwise, and this cleared off and the telescope became steady between the time given against "ingress certainly complete." There was certainly no blunting of the cusps of sunlight nor indication of the formation of a "drop" up to 0h. 23m. 47s., the time of plate II. *Apparent contact.*

At egress I watched the planet creep up to the sun's edge; there was not the slightest uncertainty or cloudiness between their limbs, and I waited patiently till I judged the sun's limb could no longer remain unbroken, remembering the thickness of the ring at ingress, before I noted the time of what I considered "real" contact under the black drop nomenclature (plate IV.) *Egress. No unsteadiness.*

Twelve seconds later gave me the appearance shown in plate V; this in the drawing is probably exaggerated, and it is the only reliable indication of a drop that presented itself; eleven seconds afterwards I judged the tangential or true internal contact took place (plate VI). *Diagram. Internal contact.*

The breadth of the ring of light is here an important consideration; during the transit I did not attempt to estimate it, but afterwards set it down as never much exceeding the $\frac{1}{100}$th part of the planet's diameter, or as 1" in thickness at the greatest. That this was an over-estimate will appear from the drawings and from the times recorded at egress. Even on the comparatively dull white of the paper a breadth of $\frac{3}{4}$" gives the full effect required, and the interval in time of 23 seconds between plates IV and VI corresponding to a motion of $\frac{3}{4}$" in the planet is a strong confirmation of the breadth given in the drawings, especially when from the nature of the phenomena the first time would probably be taken a trifle early and the second a little late, or when the planet was really slightly overlapping the sun's edge (in the drawing it overlaps considerably to give the desired effect). *Breadth of halo.*

Thus plate VI is equivalent to the estimation of the thickness of a line $\frac{3}{4}$" wide with magnifying power 96. The exact measurement is of course impossible, but that some fair approximation may be made will be evident if it is considered that a line $\frac{1}{120}$th inch thick viewed at a distance of 30 inches subtends the required angle of $\frac{3}{4}$" under such magnifying power, and a bond in the line corresponding to "the ring of light about its own thickness outside limb of sun" would be easily seen, especially to an eye practised almost daily in the reading of verniers. *Measure of $\frac{3}{4}$".*

Plate VII, fig. 2, shows the ring of light with a bright part on the N.N.E. side corresponding to plate I, fig. 3, also a decided irregularity in the shape of the planet which fig. 4, plate VII, shows to have remained unchanged for a quarter of an hour; the bulge on the south-west side does not seem to agree exactly with that shown in fig. 1 plate I, but there is sufficient approximation to make me await with interest the reports of other observers.

TRANSIT OF VENUS.

Contrast of colour. Fig. 3 plate VII is an attempt to show the contrast of colour between the disc of the planet and the sky; this colouring claimed my attention for the time, and the planet is shown as circular and the ring as uniform in brightness, simply because I have no special note about either; and my attention being otherwise engaged, recollection will not serve me further than that the ring was there and was continuous, helping me in contrasting the planet with the sky

L. A. VESSEY.

APPENDIX No. 2.
MR. HIRST'S REPORT.

Temporary Observatory, Woodford, 9 December, 1874.

I HAVE to report on matters especially under my charge and connected with the transit of Venus which has taken place to-day.

Instrument. The instrument with which I was particularly concerned was a photoheliograph by Dalmeyer, similar to those supplied by him to all the English parties, and provided with a Janson's apparatus for obtaining a number of photographs during ingress and egress. Attached to the tube of the heliograph was a finder, consisting of a single lens 1¼ in. aperture and about 4 feet focal length. This was originally arranged by the maker so as to throw the sun's image on to a piece of parchment fixed at its focus; but in order to adapt it to circumstances which required that one end of the photoheliograph should be in the photographer's dark room, the lens was inserted in the end of a brass tube, and an eye-piece provided in the shape of a Huyghenian combination, giving a power of about 50 diameters. The chromatic and spherical aberration of the single lens was in part compensated by its extreme focal length so that fair definition could be obtained of the edge of the sun and the existence of even minute solar spots made plainly visible.

To diminish the light in the finder I used a thick piece of orange-coloured glass, which gave an agreeable colour to the sun; this was placed outside the eye-lens of the eye-piece.

Ingress. As the time for first external contact drew near, I attentively watched the portion of the sun's limb where I expected Venus to appear, and within 20 seconds after first contact was reported by an observer at the 4¾-in. equatorial, I picked up the planet in the finder.

Janscn plates. I then commenced taking as many of the circular Janscn plates as possible, referring to the finder between each plate to satisfy myself that the heliograph had not shifted. Venus continued to encroach upon the solar disc, preserving at the same time a regular circular form; definition at this time excellent.

Black drop. I had prepared and placed a plate in the Janscn eye-piece, when, on taking my usual glance at the finder, I observed the disc of Venus appearing, as it wore, rather more than one-third her own diameter within the sun, and connected with the limb by a narrow line intensely black with an ill-defined edge (see plate XIV), which represents the appearance as faithfully as I can recollect; this was about five seconds before No. 5 Janscn plate was begun. I had not time for more than a glance, for I wished to procure a photograph of what I supposed to be the black drop so universally observed by astronomers more than a century ago at the last transit. On getting the plate through, however, it showed nothing of what I had so distinctly observed a few seconds before. (See plate XIV.)

TRANSIT OF VENUS.   21

Referring to the finder, Venus appeared well inside the sun, but apparently nearer the limb than she seemed before the drop was gone. I thought at the time that it might have broken before the exposure of the plate, and I determined to keep a sharp look-out for its formation at egress. Soon afterwards Mr. Vessey came in, and reported that the 4¾-in. had shown no drop at all.

Towards egress I referred constantly to the finder, that I might be ready with a plate Egress. directly the drop became visible. When Janssen plate No. 9 was in its place, and upon adjusting with the finder, I observed no black drop, the planet appearing so far within the sun's disc that I did not think it necessary to hurry in order to catch the drop, and exposed the No. 9 plate, meaning to get another in, in time. After taking out the plate, which probably occupied 20 seconds, I went to the finder, and, to my astonishment, saw that the drop had formed, Black drop. appearing about as long as one-third the diameter of the planet. I hurried on the next plate as much as possible, but a delay unfortunately of a couple of minutes occurred before it was ready; on development it showed Venus as a perfectly circular disc touching the sun's limb.

I regret exceedingly that my eye was not at the finder during the precise moment of the formation of the drop, but my duties at the Janssen eye-piece prevented me from staying there more than a few seconds at a time.

Referring to what I saw through the finder, I am convinced that my observations, short Sure of though they were, have not deceived me. I was thoroughly prepared and on the look-out for phenomena. the phenomenon at egress, and I have not the slightest doubt that any one using similar optical means would have seen what I did.

I have forgotten to mention that Mr. Vessey at the 4¾-in. reported no drop at egress.

GEO. D. HIRST.

P.S.—About mid-transit during a pause in taking photographs, I examined Venus with Halo on Mr. Fairfax's 4¼ equatorial. The planet appeared intensely black, and perfectly circular, but Venus. it was surrounded by a narrow fringe of dull red light. I was only able to watch it for a minute, when other duties demanded my attention. I used a neutral tint glass, so that the red colour cannot be attributed to it. (See plate XV.)

(Note by H.C.R.).—The exposure of plate No. 9 was begun at 3h. 51m. 42·42s. and was finished at 3h. 52m. 31·30s. At 3h. 53m. 3s. Mr. Hirst looked in the finder and saw the black drop. Mr. Vessey saw the actual contact of limbs at 3h. 54m. 48·41s. or 1m. 45·41s. after Mr. Hirst saw the black drop. So that the inferior telescope produced the black drop certainly 1m. 45·41s. and perhaps more before actual contact took place.

APPENDIX No. 3.
Mr. DuFaur's Report.

Temporary Observatory, Woodford, Wednesday, 9 December, 1874.
INSTRUMENT used, Cooke telescope, equatorially mounted, but without rack adjustment for Instrument. focus or screw motion, aperture 3 inches, power 65. Previously to commencement of transit, the only two coloured glasses having been fractured by heat of sun's rays, the field lens was

TRANSIT OF VENUS.

Focus.
Last contact.

smoked, and almost immediately succumbed to the same influence, being starred in all directions. Under these circumstances, added to the difficulty of adjustment to focus, and in retaining the sun's limb in the small portion of the field of view remaining available, all hope of satisfactory time observations were destroyed. The last contact at ingress and first at egress were observed. At ingress the whole body of the planet was discernible when only about two-thirds of it had entered on the sun's disc. I estimated an interval of about 4 seconds from the time when I judged the planet to be wholly on the sun's disc and the time

No distortion.

when the first light was apparent between their edges; though there was but little boiling of the sun's limb, this first appearance of light was by no means instantaneous, no black drop was formed, and under the circumstances in which I was placed, little (if any) distortion of the planet's following limb was observable. A gust of wind which came up at that time, together with the noise caused by the blower attached to the adjacent dark house, rendered my chronometer, which could not be placed within the range of my vision, inaudible for a con-

Perfect definition.

siderable space of time, and I could place no reliance on my estimate of the time of completed ingress.

During the time that the planet was passing over the sun's disc, between that of completed ingress and of the first contact at egress, I had full opportunities of watching its progress through the 4¼-inch telescope; the definition was very perfect at times, and I constantly focused the glass by the groups of sun-spots (of which I could detect about twenty-four) and not by the planet's disc. Thus focused, that disc appeared to me perfectly defined, at times when scarcely any boiling was observable, and was as sharp as it would be possible to illustrate it

Egress.
Lens broken.

on paper; I could detect no inequalities in its edges—it appeared to me perfectly black.

Returning to the small telescope to watch the egress I found the lens still further starred, so as to render it extremely difficult to follow the planet at all satisfactorily, but I

No black drop.

again estimated a lapse of about 40 seconds between first real and apparent contact. I again failed to observe any black drop, but the vanishing of the thread of light previous to apparent contact, so far from being instantaneous, was so gradual and ill-defined, owing to the state of my instrument and my inability to adjust the focus satisfactorily without rackwork, as my eye became weary and required such readjustment, that I am afraid I must again consider the time taken, Sydney mean time, 3h. 55m. 25.50s., to be of little value as an observation.

ECCLESTON DuFAUR.

Observations of Weather previous to and during Transit of Venus.

Temporary Observatory, Woodford, Blue Mountains, N.S.W.

Wednesday, 9th December, 1874.
    10.30 a.m.—Sky perfectly cloudless. Light westerly wind.
    11.30 a.m.—Light clouds forming.
    0.0¼ p.m.—Passing clouds crossing sun's disc.
    0.7   „ —Sun's disc obscured for about 30 seconds.
    0.9   „ —Thin clouds passing sun's disc.
    1.0   „ —Passing clouds. Sun obscured for about 5 minutes.
               Wind N.W. Almost calm at intervals.
    1.7   „ —Large cloud passing over sun.
    1.10 „ —Clear. Thunderclouds rising in the west

TRANSIT OF VENUS.

1·23 p.m.—Passing clouds.
1·25 „ —Sun's disc obscured for 10 minutes.
   Wind westerly—force about 3 with puffs.
2·10 „ —Clear. Thunderclouds to westward.
2·36 „ —Passing clouds.
2·40 „ —No material obscuration during the remainder of the transit—at its close a bank of clouds rising to south of west to within about 5° of the sun's position.

While the sun was unobscured the planet's disc was clearly visible through a simple smoked glass, up to the time when the egress was fully half completed.

Appended are hourly barometric and thermometric observations taken at Mr. Fairfax's house, Woodford.

ECCLESTON DuFAUR.

Woodford, 9 December, 1874.

Barometer and Temperature.

Hour.	Barometer.	Temp.	Hour.	Barometer.	Temp.
6 a.m.	27·88	79·0	1 p.m.	27·85	90·0
7 „	27·88	81·2	2 „	27·83	90·0
8 „	27·88	83·3	3 „	27·80	91·0
9 „	27·89	85·0	4 „	27·78	91·0
10 „	27·88	86·2	5 „	27·77	88·8
11 „	27·88	87·5	6 „	27·77	86·0
12 „	27·87	88·8			

APPENDIX No. 4.

TEMPERATURE in Clock Room at the Woodford Observatory, during the observation of the Transit of Venus, 1874.

Mean Time.	Temperature.	Remarks.
h. m.	°	
8 0 a.m.	86	
9 0	88	
9 30	92	
10 0	92	
10 30	91	
11 0	92·5	Wet blankets applied to outside of observatory.
11 30	92	
12 0 noon	{ No observation taken. }	
1 0	92	
1 30	93	
2 0	93	Wet blankets and ventilator used almost continuously.
2 30	94	
3 0	94·2	
3 30	94·3	
4 0	94	
4 30	93·8	
5 0	95	
5 30	95·9	

APPENDIX No. 5.

Results of Observations for Longitude of Woodford:—

		Difference
		m. s.
December 21, 1874	γ^2 Ceti	2 54·79
	a Ceti	2 54·79
	δ Arietis	2 54·41
	B.A.C. 1125	2 54·65
	B.A.C. 1201	2 54·60
	γ^1 Eridani	2 54·79
	o^1 Eridani	2 54·58
	ε Tauri	2 54·52
	a Tauri	2 54·50
	ι Aurigæ	2 54·41
Adopted longitude of Woodford west of Sydney...		2 54·61

	h. m. s.
Longitude of Sydney	10 4 50·81
	0 2 54·61
Adopted mean longitude	10 1 56·20 of Woodford.

Results of Observations for Latitude of Woodford:—

			° ′ ″
December 27, 1874	B.A.C. 1150		33 43 58·8
„ „ „	„ 2066		33 43 55·9
„ 28, „	„ 1433		33 43 02·6
„ „ „	„ 2066		33 43 57·6
Adopted mean latitude of Woodford			33 43 58·7

The longitude was determined by the transits of stars over the Sydney and Woodford transits, the times being recorded on the Sydney chronograph.

The latitude was determined by placing the transit instrument in the prime vertical.

APPENDIX No. 7.
Mr. Fairfax's Report.

Woodford, 10 December, 1874.

Halo. During the time when the disc of Venus was slowly creeping on to the sun, I had the opportunity of using my own glass (4½-in. equatorial) for a few moments. Venus was at the time about two-thirds on the sun and appeared to me perfectly black, and all that part which was seen with the sun as a background seemed to be as sharply defined as possible without any haziness, but with the part off the sun it was different; that was marked by a most brilliant line, which was very narrow, and yet seemed to me like my drawing, plate XXIX, which represents the general appearance, but I would not for one moment say that the drawing is correct to scale; in fact the halo was so narrow that I could not make a drawing

Colours. like it. Still I distinctly saw in that very narrow halo such forms and colours as I have represented. It appeared as if Venus were surrounded with white and red flames mixed and so close together that they formed a continuous ring which was probably less than $\frac{1}{16}$ of the diameter of Venus across.

A. FAIRFAX.

CAPT. HIXSON'S REPORT.

Goulburn, 10 December, 1874.

THE extraordinary power of the sun and its vertical position during the period of ingress interfered materially with my observations. The weather was comparatively mild previous to the 7th of the month, but on that day the maximum thermometer reached 102° in the shade, the wind being from the westward and the atmosphere slightly disturbed by electric storms. A small quantity of rain fell, but not sufficient to make any appreciable show in the rain-gauge. On the following day, the 8th, the day previous to the transit, the weather was of the same character, the maximum thermometer registering 103° in the shade. _{Power of sun. Temperature.}

The 9th, the day of the transit, set in with the wind from the westward, light in the morning, with occasionally drifting clouds; the wind increased to a force of about 6 in the Beaufort tables by 2 p.m.; at this time about half of the sky clouded, the clouds occasionally but not seriously interfering with the observations. The thermometer at noon stood at 101°; at 1 p.m. it was 102°, and at 2 o'clock it was 104° and it remained at this point up to 5 o'clock. The thermometer in the sun was not regularly recorded, but at noon I am informed it was observed to be 130°. _{Weather of 9th. Strong wind. Clouds. Temperature.}

The barometer was 27·82 inches at noon, 27·79 inches at 2 p.m., and 27·73 inches at 5 p.m. _{Barometer.}

The heat was so great in the observatory for photographic work, notwithstanding that Mr. Russell had fitted a ventilating fan to it, that both I and Professor Liversidge remained in the sun uncovered some seconds after coming out without being conscious of the danger we were incurring until our attention was called to the fact of our being bareheaded by a person in attendance. _{Great heat. Ventilator.}

I used a 6-inch equatorial telescope with a magnifying power of 130 diameters. A shaded glass had been broken by the power of the sun the day previous, and therefore I was careful to focus the instrument in such a manner as to expose my shades to a gradual heat. Notwithstanding this care but a short time elapsed before the glass in front of the lens became fractured and it had to be removed. This was replaced by a smoked glass, but soon the smoke evaporated and the glass became so cracked as to _{Instrument. Glass broken.}

D

Coloured shades.

be altogether useless. I next, with the assistance of Mr. Tornaghi, tried a smoked glass over the object-glass of the telescope, but now the coloured shades in my movable eye-piece both gave way, so that I was reduced to holding in my hand the shaded glasses which Captain Onslow permitted me to take out of his sextant. These became so heated as to be difficult to hold; this, coupled with the uncomfortable position I had to assume, induced me to take Mr. Tornaghi's suggestion and attach the camera to the telescope, and watch the image on the ground glass. In this way I first observed the planet on the sun's disc at about 12h. 2m. 41·00s. The planet as I had the opportunity of observing it through the telescope appeared to work its way on to the sun's disc without becoming disturbed or to assume the shapes which we had been led to expect. The edges of each object were well defined, the planet appearing quite dark and globular.

Contact.

Ingress.

First internal contact was clearly noted to break by Mr. Tornaghi observing the light of the cusps forcing its way from either side between the planet and the sun at 12h. 25m. 50·27s.

Photographs.

Photographic work was commenced after first internal contact had taken place.

Egress.

My observations of egress were taken by holding a shaded glass to protect my eyes, all the coloured glasses of the telescope having become fractured as before described; but this became an easy task now, as the position of the sun was much more favourable for observations than it was at ingress, and it had moreover lost much of its power.

Globe.

At 3h. 46m. 4·38s. the planet appeared a dark beautifully defined globe, the sun's edge being also sharply defined.

First contact.

I judged that contact took place at 3h. 54m. 28·01s. Previous to this there was a portion of light visible between the planet and the sun.

Bisected.

At 4h. 10m. 0·45s. the planet appeared to be bisected by the edge of the sun, both objects being clearly and sharply defined. The portion of the planet off the sun was invisible to me.

Several clouds passed over at about this stage of the phenomenon.

Last contact.

At 4h. 23m. 34·24s. I recorded last external contact, after which the planet disappeared altogether. I observed a portion of the planet illuminated, as I thought, at one time, but I do not consider my observations in this respect of any value, as my shaded glasses were held in my hand and occasionally spoiled the focus.

The day after the transit was not nearly so hot, and the day after that the weather became quite temperate, and cold towards evening.

Appended will be found Reports from
Capt. A. Onslow,
Professor Liversidge,
Mr. A. Tornaghi.

THE instrument at my command for determining the latitude and longitude of the station was an altazimuth. Telescope 1¾-in. aperture and focal length 19-in. vertical circle 15 inches provided with two micrometer microscopes reading to 1". *Altazimuth.*

As soon as this instrument was placed in the meridian the azimuth circle was clamped and the observations made as with a transit circle. For the determination of the difference of longitude, selected stars were observed in transit over the meridian of Sydney. I then observed them in transit over my instrument, and by means of a contact key sent the time of transit over each of the seven wires to the chronograph in Sydney Observatory; the difference between the transit at Sydney and Goulburn was thus measured by the standard clock, each star afforded an independent determination of the difference in longitude, and it is satisfactory to find such a small range in the differences, when the size of my instrument is considered. *Longitude and latitude.*

I used the same stars together with a few others to determine the latitude, the observations being made in the same way as with a transit circle. The differences are not so small as I should have liked to see them, but no doubt the extreme heat of the weather gave rise to the difficulty experienced in bisecting them.

The observing tents were placed in the middle of the Market Square, Goulburn, and the brick pier on which the transit instrument was placed was left standing when we left.

The height of the observatory above mean sea level was 2,129 feet.

RESULTS of Observations for Longitude of Goulburn :—

		Difference.
		m. s.
December 11th, 1874...	δ Arietis ...	... 6 0·32
	η Tauri ...	... 6 0·25
	No name	... 5 59·90
	ε Tauri ...	... 5 59·98
	α Tauri ...	... 6 0·34
	α Tri. Aust....	... 6 0·13
	ι Aurigæ ...	... 6 0·27
	ε Leporis ...	... 6 0·04
	β Orionis ...	... 6 0·02
		6 0·14

	h. m. s.
Sydney longitude	10 4 50·81
Difference of longitude ...	6 0·14
Adopted mean longitude ...	9 58 50·67

TRANSIT OF VENUS.

RESULTS OF OBSERVATIONS FOR LATITUDE OF OBSERVATORY.

December 11th, 1874...	θ Ceti		34 44 43·70
	η Piscium	...	34 45 4·60
	a Eridani	...	34 44 41·50
	a Arietis	...	34 45 24·30
	ζ² Ceti	...	34 45 9·80
	δ Arietis	...	34 45 52·00
	η Tauri	...	34 45 45·20
	o¹ Eridani	...	34 45 21·10
	ε Tauri	...	34 45 36·90
	a Tauri	...	34 45 0·20
	ι Aurigæ	...	34 44 38·50
	ε Leporis	...	34 44 56·80
	β Orionis " Rigel"	...	34 45 29·30
Adopted mean latitude	...		34 45 12·68

FRANCIS HIXSON,
Capt., R.N.

CAPT. ONSLOW'S REPORT.

Ingress.

Telescope used 3¾-in. equatorially mounted refractor.

THE time of the planet's first apparent contact not recorded, as the planet was not seen till it was well on the disc of the sun.

At 12h. 10m. 14·24s. half the sphere of the planet apparently on.

At 12h. 13m. 54·25s. the planet somewhat resembled the letter D or the top of a thumb projecting over the sun's surface.

At 12h. 16m. 28·26s. a bright light was seen at the left point of intersection of the two circles (fig. 1, plate XVI.), and in a few seconds a similar light at the right point (fig. 2); time 12h. 16m. 34·26s.

At 12h. 19m. 29·28s. an apparent circle formed by planet, 12h. 21m. 20·28s. Venus apparently just touching inner edge of sun.

At 12h. 23m. 29·28s. the internal contact appeared complete, but at this moment the objects got out of my field of vision, and when again sighted 12h. 25m. 29·28s. the planet was well inside the sun's limb.

Egress.

Observed by Mr. Tornaghi, time recorded by me :—

At 3h. 53m. 47·78s. the light between the internal edge of the sun and external edge of the planet was a little dim though the circles were quite distinct.

3h. 54m. 25·79s. time of contact.

3h. 58m. 44·80s. about ¼ off.

4h. 3m. 14·82s. the upper intersection to the right a little flattened.

Planet about half on 4h. 8m. 50·80s.

about one fourth on 4 18 22·83

disappearance ... 4 23 27·84

A. ONSLOW.

Professor Liversidge's Report.

Goulburn, 9 December, 1874.
Temperature in shade, 104·4.

The instrument used was a 3¼-inch equatorial telescope, and the power employed was 150 diameters. A red dark glass was screwed on over the eye-piece in the ordinary way; the neutral tint and blue dark glasses attached to a slide were found to be less convenient for use although their lights were softer and less trying than the red. *Instrument.*

The telescope was placed upon the brick pier built for the transit instrument; this pier was about 2 feet square, and rose some 22 inches out of the ground; and the only position which was permitted me during the ingress was a reclining one, with my feet to the east and my head to the west. I may mention that it was an uncomfortable and unsteady one, and prevented me from taking such full notes at the time as I had wished.

The first stages of the ingress were not observed, so I accordingly waited until I judged the planet was half on before recording any observations; this took place at 12h. 7m. 0·6⅓s.; the planet was apparently half on or bisected. *Bisected.*

Ingress.

At 12h. 20m. 5·14s. I was inclined to think that there was "apparent internal contact," as seen in the artificial transit. *Internal contact.*

At 12h. 21m. 4·14s. I again judged that there was "apparent internal contact," and considered that my estimation of it at 12h. 20m. 5·14s. was made too soon.

At 12h. 22m. 5·6⅓s. the circle of Venus appeared to be complete, and apparently just touching the sun's limb, i.e. "internal contact" really took place. *Complete ingress.*

At 12h. 25m. 39·15s. the disc of Venus was clear of the sun's limb, and appeared to be about ⅓ of the planet's diameter within it.

During the interval between the times 12h. 22m. 5·6⅓s. and 12h. 25m. 39·15s. a faint hazy grey filament like a streak of smoke was momentarily observed between the edge of the planet and the sun; it was very obscure and ill-defined. *Haziness.*

I unfortunately failed to note the exact time at which the cloudiness was present between the two limbs, for while trying to get it more in focus and more sharply defined it vanished. There did not appear to be any sudden break in it, but it faded away quite imperceptibly.

No traces of the "black drop" were seen, unless the above be considered such.

The absence of the black drop and the unexpected manner in which the planet made its ingress, unaccompanied by the distortions and other peculiarities previously predicted, rather upset my expectations, and tend to make my observations of this portion of the transit less comprehensive and detailed than they might otherwise have been.

Egress.

At 3h. 39m. 40·31s. the planet was about one-third its diameter from the sun's upper left-hand limb; it then appeared spheroidal, and not as a disc merely; it appeared illuminated on the inner side in the direction of the sun's diameter, and this illumination *Spheroidal.*

shaded off on each side of the planet, but at the portion nearest to the sun's limb it appeared quite black and opaque (Plate XXII, figs. 1, 2, 3).

Globular. This globular appearance was retained until the planet had passed off the sun's limb to the extent of about ⅓ of its diameter.

Haziness. At 3h. 46m. 40·31s. I fancied I could see a slight haziness between the planet and the solar limb. I do not attach any importance or value to this observation, as the haziness was exceedingly ill-defined.

Contact. At 3h. 54m. 57·31s. contact between the two limbs took place.

Processes. At 4h. 0m. 2·31s. the planet was just beginning to pass off the sun's limb, and it looked somewhat as if it were pushing that portion of the sun's limb before it, for the solar limb appeared to be raised up into two processes, one on each side. (Figs. 5 and 6, a and b.)

At the time I thought it might perhaps be due to an atmosphere surrounding Venus, or to an optical illusion; but since I have heard that other observers saw the illumined edge of Venus beyond and outside of the sun, I am inclined to think it was that which I saw. I, however, did not see a segment of a circle beyond the sun, but merely two portions or cusps brightly illuminated, but not as bright as the sun.

The cusps of the sun around Venus appeared brighter than the body of the sun.

D shape. At 4h. 3m. 10·31s. Venus appeared to be nearly ¼ off the sun's limb. There was just the slightest trace of distortion or tendency to the D form retained until the planet was half off, but hardly perceptible.

At 4h. 11m. 15·31s. the planet was half off; at this stage and afterwards there was not the slightest traces of distortion.

At 4h. 10m. 40·31s. three-quarters off.

At 4h. 23m. 30·31s. there only remained the slightest indentation to mark her presence.

At this moment a cloud passed over the sun from left to right, and at 4h. 24m. 2·31s. all traces of the planet disappeared from the sun's disc, *i.e.* final external contact at egress took place.

ARCHD. LIVERSIDGE,
Professor of Geology,
Sydney University.

Goulburn, 10 December, 1880.

Mr. Tornaghi reports as follows:—I took the time when the line of light between Venus and the sun's limb at egress disappeared as the time of contact.

Halo. After this I saw the halo, and it was best when the planet was half off the sun. The outside had a greenish colour with red in it, and appeared as if formed by flames issuing from the planet all round, and densest at the planet. The halo round the part on the sun was different, but quite distinct and unmistakable.

A. TORNAGHI.

REPORT BY THE REV. WM. SCOTT, M.A., ON THE TRANSIT OF VENUS
AS OBSERVED AT EDEN.

WE left Sydney on Tuesday, November 21th, and arrived at Eden, Twofold Eden. Bay, the next morning. Having landed our observatory, tents, and instruments, together with a good supply of bricks and cement for building piers for the instruments, my first care was to find a suitable spot for the observatory. I was not long in selecting an open space known as the Market-square, on a hill overlooking both bays. This site has the advantage of being near the telegraph line, and commanding uninterrupted views of the ranges at some miles distance to the south and west, the wooded sides of which I saw would afford good reference marks for the adjustments of Instruments. the transit instrument in the meridian and prime vertical. The day was nearly over before we had carted all our baggage to the top of the very steep hill which forms the principal street. A commencement however was made of setting up the observatory, in which we were most effectively assisted by Mr. Russell, the Harbour Master, and his boat's crew. On Saturday everything was ready, with the exception of mounting the equatorial telescope, which was delayed in order to allow the pier to become quite dry. An approximate meridian had been determined by sun observations with a theodolite.

Our instruments were—the $7\frac{1}{4}$-inch equatorial telescope formerly used in the Sydney Observatory (with good driving-clock), a portable 2-inch transit instrument, a $4\frac{1}{4}$-inch equatorial telescope and a $3\frac{1}{4}$-inch equatorial telescope, the theodolite before mentioned, a clock and three chronometers. The upper portion of the observatory was provided, besides the usual shutter, with a frame fitting the opening, to which was attached a bag of yellow calico, of somewhat conical form, having a hole in the smaller end, through which the telescope and finder could pass. This bag being secured round the middle of the telescope tube, excluded all but yellow light, so that the whole observatory answered the purpose of a dark room for photographic work.

This arrangement, though very convenient, was liable to be influenced by the wind, and so to interfere with the steady motion of the telescope. All being ready, I waited anxiously for a clear night to enable me to make the necessary star observations for time and instrumental adjustments; but so unusually cloudy was the weather that I could get no satisfactory observations until Saturday, December 5. On the 7th and 8th I exchanged longitude and clock signals with the Sydney Observatory, but on each occasion was prevented by clouds from getting more than one transit observation.

On the 7th, being a clear day, we took two sets of photographs, to satisfy ourselves that all was in good working order, and found that by reducing the aperture of the telescope to 3 inches the sun's edge was more sharply defined and the reference lines more clearly distinguished.

In our trial observations of the sun several of our dark glasses were cracked by the heat; so finding that I could get no sufficient protection even with the 3-inch diaphragm, I constructed one 2 inches in diameter, which gave very satisfactory results.

Cloudy. In consequence of the continued cloudy weather my instrumental adjustments were not so accurate as I wished. I was assisted one day in adjusting the reference lines in the camera by a small well-defined solar spot, which appeared to traverse one of the lines with great accuracy. In order to correct any remaining error in the position of the lines, I adopted the plan recommended by Mr. Russell of taking two photographs of the sun, at an interval of about a minute, on the same plate, and determined to repeat the process at every half-hour during the transit. Now, in order to make this double image of any service, it is necessary that the common tangents to the two images should be exactly parallel to the direction of the sun's motion. For this purpose the telescope must remain perfectly at rest, and therefore must not be touched during the interval.

Flashing shutter. This result appeared difficult to obtain, as the flashing shutter must be made to cross the field a second time for the second image. The method which I contrived, though somewhat complex, appears to be perfectly satisfactory.

The flashing shutter, as described by Mr. Russell in his paper read before the Royal Society on September 3, 1873, is attached to the end of a lever, which is drawn down by an elastic band, when the other end is released by pressing a spring. If the second image were obtained by raising the shutter quickly by the hand at the end of the desired interval, the action of so raising it would probably displace the telescope; or if the dome shutter were closed, or a cap placed on the telescope, and the flashing shutter restored to its former position and again released, there would be the same risk and almost certainty of displacement. To overcome this difficulty I arranged as follows :—For distinctness I call the end of the lever to which the shutter is attached A, and the opposite end B. I attached a piece of wood to the camera so as to project over B. An elastic band secured to the camera below, and enclosing B, was tied by a string to this projecting piece, so as to allow B to move freely within it. When a double image is to be taken, the telescope is so adjusted, by the help of the finder, that a little more than half of the sun shall appear in the photograph, The driving-clock is then stopped and the photograph taken in the usual way. The telescope remains at rest for a minute; meanwhile the band which pulled down the flashing shutter is cut with a sharp pair of scissors, and at the end of the minute the string which holds the band at the end B is cut; B is thus drawn down and A flies up with the flashing shutter, so that a second image is taken. As an elastic band is cut each time, it is necessary to have as many bands round the camera at A, and as many loops of string at the piece over B, as there are double images to be taken.

On the morning of the 9th the weather seemed promising. I ^{Weather.} obtained clock signals from the Sydney Observatory, and by 11 o'clock we were all collected and anxiously waiting for the transit to commence. Clouds were coming up and the wind rising, and we had reason to anticipate a disappointment. At the time of ingress, however, the clouds had not yet intervened. The exact instant of first contact it was impossible to determine. Mr. MacDonnell recorded 11h. 56m. 29s. Sydney mean time as the moment at which he became quite convinced that the transit had commenced. I found my 2-inch aperture answer admirably, not only from the diminished light and heat, but also from the great distinctness of the outlines of the sun and planet. I soon became convinced that all we had heard and read respecting the apparent elongation of the planet's disc, and formation of what has been described as the "drop," was a delusion. For some minutes before internal contact I could see clearly the whole of the planet's outline; in fact, it presented exactly such an appearance as might have been expected from a planet possessing an atmosphere. Whilst the direct light of a portion of the sun was shut out by the intervention of the planet, a sufficient portion of that light reached the eye by refraction, through that atmosphere, to render the whole outline visible. By means of a double-wire position micrometer, I obtained a measurement of the apparent diameter of Venus; then, bringing one of the wires into the

F.

position of a tangent to the sun's limb, waited until the planet seemed to touch the other wire. This occurred at 0h. 21m. 7s., though Mr. Mac-Donnell, who judged the same phenomenon by the eye, unaided by a micrometer, placed it nearly two minutes earlier, or at 0h. 19m. 24s. This I

Ingress. believe to be the most important determination, being the moment of complete ingress; and I regret that the action of the wind on the telescope rendered it impossible to keep the micrometer wire in its true position as a tangent to the sun's limb. Still, I consider the above result to be very near the truth. I continued to watch the planet for more than three minutes, and saw the partial obscuration of the sun's limb by the planet's atmosphere gradually diminishing until it disappeared altogether, when I left the telescope at 0h. 24m. 48s. Mr. MacDonnell's estimate of the same phenomenon was 0h. 25m. 14s. The discrepancy between Mr. MacDonnell's results and my own shows how impossible it is to fix the moment of a phenomenon of the kind, when the motion is so slow and the change from darkness to light so gradual. The slow rate of the planet's motion across the sun's disc may be estimated by considering that it occupied over four hours in describing so small an arc, not far exceeding one-half of the sun's diameter. The difficulty was still further increased by the planet's path not being at right angles to the sun's limb, but inclined to it at an angle of about 32 degrees.

As soon as we had concluded that ingress was complete, the 3-inch diaphragm was substituted for the 2-inch, and we proceeded to take photo-

Photographs. graphs, but in doing so we were very much impeded, and the quality of the pictures affected by the clouds which were continually driving over the sun's face: indeed there were very few minutes during which the sun was not more or less obscured. Again, the action of the wind on the yellow bag was so great that the driving-clock became almost useless, and I was obliged to hold the telescope as best I could, with my eye at the finder, whilst the plates were inserted and the flashing shutter released. We made two attempts at a double image, but of course the results were quite unreliable. On the whole we took about fifty photographs, very few of which I fear are of any value. At one time we had to stop for twenty, and at another time for eighty minutes, the sun being entirely obscured. On the whole the expedition to Eden has not been so successful as I wished it to be, and I came away under the impression that Eden, though a beautiful spot, and in many respects a most desirable place to inhabit, is about the worst place for astronomical observations that I ever visited.

The longitude of the Eden observatory, as determined by transits of two stars over the meridians of Eden and Sydney, the times being recorded on the Sydney chronograph. The weather continued so cloudy and unfavourable during my necessarily limited stay at Eden, that I could not obtain transits of more stars, and am reluctantly compelled to base the longitude upon the following differences of longitude :—

	h.	m.	s.	
	0	5	11·01	west of Sydney.
	0	5	11·13	,,
Mean	0	5	11·07	
Longitude of Sydney	10	4	50·81	
Longitude of Eden observatory	9	59	39·74	

Latitude and longitude.

The latitude depends upon observations made with the transit instrument in the prime vertical, and the mean result is $37° 3' 47''$ south.

W. SCOTT.

NOTE.—Mr. Scott returned to Sydney overland, and during the long and troublesome journey he unfortunately lost all the papers connected with the determination of the position of the Eden observatory. I am therefore unable to give the separate observations for latitude.

H. C. RUSSELL.

MR. MACDONNELL'S REPORT.

Eden, Twofold Bay, 14 December, 1874.

H. C. Russell, Esq., F.R.A.S., &c.,
    Government Astronomer for New South Wales,—

  Sir,

    I have the honor herewith to forward my report of observations of the transit of Venus, as seen by me at Eden, Twofold Bay, N.S.W., on the 9th December, 1874.

The telescope entrusted to my charge was an achromatic equatorially mounted by Cook & Sons, of York, clear aperture $4\frac{1}{4}$ inches, focal length 60 inches. The eye-piece used gave a power of 98 diameters (marked 100 by the makers), and the sun's light and heat were modified by a diagonal reflector of unsilvered glass, thus enabling the full aperture to be effective, and a very light bluish screen was all that was necessary for the protection of the eye.

Instrument.

Weather.	In the early morning the weather was fine, giving our party promise of successful observations, but later in the day the sky clouded over. Towards noon it cleared up again overhead, and the observers took up their posts. I used a chronometer, No. 419, maker, Hornby, of Liverpool.
First contact.	A little after 11h. 57m. Mr. Scott called out that he saw Venus entering on the sun. I did not perceive it till about 11h. 57m. 30s., when the planet was fairly encroaching on the sun, appearing like a small notch cut out of that body.
Halo.	The planet continued slowly to advance, and 12h. 4m. 59·8s. was noted by me as the time of apparent bisection; a shadowy nebulous ring seemed to envelop Venus on the preceding side; it was of lighter tint than the planet, but was decidedly perceptible, and appeared to be about a quarter or a fifth of Venus's diameter in width.
Halo.	When the ingress was about two-thirds completed the whole outline of the planet was distinctly visible in the telescope, the shadowy envelope surrounding it very plainly. (See plate XVIII.) Perhaps it was the solar atmosphere that served as a background to throw the planet out into relief; whatever was the cause, however, the phenomenon was easily seen.
Internal contact.	Apparent internal contact was noted at 12h. 19m. 24·2s., and all attention was now devoted to the formation and breaking of the "black drop." As Venus proceeded, the
Haziness.	shadowy envelope disappeared except between the planet and sun's limb, where it seemed to fill up the space between them with faint rings concentric with the planet's edge. There was no distinct rupture of this appearance, the light seeming to go in and out several times, and
Ingress.	prevented any accurate determination of the completed ingress. I marked 12h. 25m. 14·7s. as the time, but feel now convinced that it took place at least 15s. earlier. Mr. Scott's determination was 24s. earlier than mine, but he thinks he was a little too soon, as the whole phenomenon was too indistinct to be noted accurately. There was no abrupt breaking of the "ligament," if it can be so called, but a gradual dissolving away.
Halo gone.	Ingress being now completed, the camera was fixed on the large equatorial, and some fifty photographs of the transit taken. Clouds thick and heavy covered the sun, completely putting an end to our operations. We had some momentary glimpses of the transit through the telescopes, and I noted the complete disappearance of the envelope already referred to the outline of the planet was very sharp and distinct, like a hole bored through the sun.

A heavy black cloud once more impeded our view, and egress was not observed at all much to our disappointment.

The times of the various phenomena noted above are as near as I could judge them but I cannot place much reliance on them, as there was so much difficulty in their determination.

I have, &c.,
W. J. MACDONNELL.

Mr. Watkins' Report.

Eden, Twofold Bay, 10 December, 1874.

To H. C. Russell, Esq., Observatory, Sydney,—

Dear Sir,

As a member of the expedition sent to Eden, Twofold Bay, to observe the transit of Venus of 1874, I have the honor to send you my report.

I regret that I missed observing the respective times of the external and internal contacts on the ingress of the planet. Observations of the egress were prevented by dense clouds entirely obscuring the face of the sun; I can therefore only report of my observations of the phenomena accompanying and immediately following the ingress.

Instrument. — The instrument used by me was an achromatic refracting telescope, of $3\frac{1}{2}$ inches aperture, stopped down to $1\frac{1}{4}$ in. The eye-piece was direct, inverting, magnifying about 120 times, and fitted with a dark glass.

Contact. — I first saw the planet some half minute after the external contact was observed by the Rev. Mr. Scott and Mr. MacDonnell. The edge of the disc formed by the planet appeared to me clear and sharp. I did not see any halo; but as I did not observe the planet with the intention of noting phenomena other than the times of contact, a faint halo might well have *Halo.* been observed by others without being noticed by me.

Elongated contact. — As the internal contact drew near the planet seemed to cling to the edge of the sun and so adopt a slightly, but very slightly, elongated or oval form. (Plate XII.)

Just before the time when, from the observations of Messrs. Scott and MacDonnell, the internal contact took place, I observed very thin lines of light flash in a direction parallel to the edge of the sun, in the dark broad neck joining the planet with the edge of the sun. The remarks I have before made with respect to a halo must in part be applied to what I saw of the lines of light. I can only say that I observed the lines of light, and cannot be positive one way or another as to the direction in which the flashes appeared to move, or whether or not they appeared instantaneously and vanished in like manner.

The actual moment of contact I did not see, but very shortly after the time when Mr. MacDonnell said that contact was made I saw the planet well within the disc of the sun.

As I have before mentioned, no observations were taken of the egress.

I am, &c.,
JOHN L. WATKINS.

GENERAL REPORTS.

DR. WRIGHT'S REPORT.

Weather. The morning of the 9th of December, 1874, at daylight, was calm, a heavy mist obscured distant objects, and for some time after sunrise was so dense that the sun could be gazed at with the naked eye. About 8 a.m. the sky began to clear, and it became evident that the day would be fine and intensely hot. It was my intention to take the time of the internal contacts at ingress and egress only, and during the rest of the transit to observe any phenomenon that might present itself. To assist me in this, Mr. Russell rated my chronometer at 11·30 a.m. by the sidereal clock of the Observatory, and this was again done when the

Instrument. transit was over. The telescope used was an 8¼-inch Browning, with silvered glass reflector of 68½ inches solar focus, provided with one of Browning's solar eye-pieces (having two prisms arranged for single reflections). To this was attached a positive eye-piece by Ross which gave a magnifying power of 175 diameters. The only shade required to protect the eye with the solar eye-piece was a light smoke-coloured glass. This was free from heat when exposed to the full aperture of the telescope. For observing the contact at ingress and egress the aperture of the telescope was reduced by a stop to 5¼ inches. In some other

First contact. observations the full aperture was used. At 11h. 55m. 22s. the first contact of Venus and the sun's limb took place, 3 minutes and 4 seconds later than the time given in the Nautical Almanac. Definition of both sun and planet was perfect, and the margin of each was entirely free from colour. When Venus had made a perceptible notch on the sun's edge I looked particularly for the outline of the body of the planet, but it was undistinguishable

Halo. from the black background of the sky. When Venus was nearly half on the sun, I noticed a slight form around that portion of the planet yet off the sun's disc. This brightened every moment, so that in a very few minutes it presented a bright line of light around the planet's edge, throwing it out in bold relief against the sky and giving Venus a stereoscopic appearance. The planet now looked to me like a black ball suspended in the sky, that portion of the disc which was on the sun being intensely black, whilst the remainder of the disc off the sun and near the halo appeared decidedly lighter. The portion close to the halo was shaded with reddish-brown colour. As Venus passed slowly on to the sun's disc, the bright halo which at this time was about 1" in diameter became very bright, and was observed until Venus was fairly on the sun.

Black drop. Warned by a statement made by the French astronomers that very possibly the "black drop" might be absent, my attention was wholly taken up by watching the phenomena at

Good definition. internal contact. The margin of Venus' disc continued sharply and beautifully defined as it passed on to the sun. A slight shimmering of the solar edge might be observed at the

Haziness. moment when the two outlines as it were of the sun and planet touched, and also some slight shading of the planet's edge (something like a penumbra), but it was clear to me that there was no "black drop," nor any such elongation or distortion of the black edge of Venus that could be taken for it. I was so intent upon observing this that I allowed some seconds to elapse before I recorded an observation of the internal contact at 12h. 24m. 30s.; this I have no doubt was half a minute after actual contact, and that it was so is proved by Mr. Russell's observation at 12h. 23m. 59s.

The planet was now fairly on the sun, the halo had entirely disappeared, although carefully looked for with the full aperture of 8½ inches, nor could any irregularity of the edge of the planet be detected. For an hour Venus was constantly watched in her path across the sun. Her neighbourhood was closely scanned to find if possible any small speck which might denote the existence of a satellite, but with negative results. No fresh phenomenon was seen. Urgent professional duties called me away for nearly an hour and a half, but I was able at a quarter before 3 o'clock to renew my observations. Meantime the heat of the day had increased to 87° F. in the shade, and the black bulb thermometer in vacuo showed 124·5 F. in the sun. The breeze from the N.E. had freshened and was laden with moisture which caused a slight haze in the sky, and at times some unsteadiness in definition, which, however, on the whole remained sharp and good. Venus on the sun.

The planet was approaching the N. by W. edge of the solar disc, and when viewed with a low power (40 diameters) appeared as a perfectly black spot on the sun; with the full aperture and a high power (250 diameters) the outline of Venus was still free from any appreciable irregularity.

Determined to take the most accurate observations possible of the internal contact at egress, I carefully set my chronograph by the chronometer, and was fortunately enabled to catch the precise moment of internal contact at egress, at 3h. 54m. 30·59s., Sydney mean time; the edge of the planet coming at that instant sharply and distinctly in contact with the sun's limb, it was quite as clear as at ingress that there was no black drop. The accuracy of this observation was corroborated by Mr. Russell at the Sydney Observatory, which is 2,300 feet north and 792 feet west from my house. His time for this contact was 3h. 54m. 30·66s. He was using the new 11¼ refractor of the Observatory, of 12 feet 6 inches focus, with the aperture contracted by a stop to 6 inches, and an eye-piece magnifying on his telescope 100 diameters. The edge of the planet which was in contact with the sun's limb now was observed to assume a square form, from blunting or rounding off of the solar cusps. This lasted a very short time, perhaps half a minute. (Something of the same appearance was observed at ingress before internal contact, but it was so slightly marked that it did not excite any attention at the moment.) When Venus had still further passed off the sun, I noticed the reappearance of the halo around the dark body of the planet again, throwing that portion of her disc (as at ingress) into relief against the sky. This halo gradually became brighter, and was not uniform as at ingress, but most distinguishable on the N.E. quadrant of the planet; here it presented a decided accumulation of light, especially about the centre of the quadrant, and at this point encroached a little upon the dark outline. (See plate XXIII.) The full aperture of the telescope was used, and showed the colour of the halo and the solar light to be the same, and gave the impression that the increased light at the spot above mentioned was due to reflection from polar snow. The same appearance of shadings of rusty brown colour was observed at the margin of Venus' disc as at ingress, and as she passed onwards from off the sun the halo gradually faded until she ceased to be visible, after the last contact at 4h. 24m. 27s., forty-five seconds before the predicted time. Some very fine groups of spots were seen on the equatorial zone of the sun, but as they were far south of the part of the transit no particular observation was taken of them.

H. G. A. WRIGHT,

Mr. Allerding's Report.

Dear Sir,   Sydney, 16 December, 1874.

I have much pleasure in handing you my report of the transit of Venus as observed by me.

External contact. The external contact at ingress I saw well defined at 11h. 50m. 6s., but at the internal contact at ingress at 12h. 24m. 14s., I saw first a haziness between the limbs (plate XIII, fig. 3), and this turned into a cone (fig. 2), and when it had nearly disappeared it seemed to stretch out to a fine thread (fig. 1), by which Venus seemed to be attached to the edge of the sun—it seemed as long as ⅛ of the diameter of the planet—and then this line instantaneously disappeared at 12h. 24m. 44s., but Venus was then already well detached from the sun's limb. Had I not waited for the disappearance of the fine line I would have made the inner contact at least 30s. sooner.

Egress. The internal and external contact at egress I cannot be very certain about, having had so many interruptions by having to allow a great many friends to have a peep at the transit, but I give the time as near as possible at internal contact, 3h. 54m. 35s., and external contact 4h. 23m. 48s.

Instrument. I am surprised that no one in the Observatory saw anything like a drop, and my telescope defines well, for every one that saw Venus on the sun's disk remarked its clear and sharp definition. But I must draw your attention to my having put a cardboard cap over the object-glass, with a 2-inch aperture, to get rid of the heat on the eye-piece. The glass I was using has an aperture of 3½ inches, but it was not well placed, as I was obliged to use it in my back yard which is surrounded by buildings.

                                                           I remain, &c.,
                                                              F. ALLERDING.

P.S.—My house is situated in Hunter-street, 1,980 feet south and 1,452 east from your Observatory.   F.A.

Mr. Bolding's Report.

                                          Raymond Terrace (3 miles west of Newcastle),
                                                                     10 December, 1880.

Instrument. For the purpose of observing the transit of Venus I provided myself with a marine chronometer, which was carefully rated at Newcastle, by means of the time signals sent from the Sydney Observatory to drop the time-ball; I hope therefore my time may be relied upon.

My telescope was a very good 3-inch refractor, stopped to 1⅞ inch, equatorially mounted. To help in securing exact time I got the services of a friend, who noted the times and any remarks descriptive of phenomena which I made.

At the commencement the sun's edge was very unsteady and seemed to be "boiling," Boiling. and I did not catch the moment of first contact; when I saw it there was a very perceptible dent in the sun's limb; five minutes afterwards the planet seemed to be half on the sun, then appeared for an instant a tendency to a straightening of the curve, but at 12h. 13m. 40s. there appeared a distinct shoulder on the north side (i.e. as seen inverted on the south, see fig. 1, Shoulder. plate XXVIII). The rim was as dark as the planet itself, but unsteady, and better defined towards the north than on the opposite point of the planet. This appendage seemed to shrink up as the planet crept wholly on to the disc after these shoulders disappeared, and Venus came Shoulders on with great steadiness, but at the moment I expected the complete circle came the form disappear. shown in fig. 3, which at the time I called a parachute, time 12h. 30m. 12½s.; upon this came a haziness which I have shown in the same figure, and at 12h. 23m. 43⅜s. I noted complete ingress. Haziness.

Egress.

As the time of egress came on the definition was very good indeed, and the sunlight was Good defini- now sufficiently reduced, by the sun sinking to the west, to enable me to use the full aperture tion. of my telescope, and I saw the planet make internal contact at 4h. 0m. 3·5s. without any of the peculiarities noted during the unsteady definition at ingress. At 4h. 7m., when examining the points of contact or cusps, I saw a silvery line of light extending partly round the west side Halo. of the part of the planet off the sun (fig. 4); 30 seconds later I saw it all round that part of Venus as in fig. 5, and I continued to see more or less of this beautiful silvery line until 4h. 11m., when I saw the last of it on the west side; it looked like a silver edging as if caused by refraction from an atmosphere.

The planet then passed off with nothing more remarkable than an occasional blunting of the cusps from atmospheric disturbance, and at 4h. 20m. 7s. I noted last external contact.

The following notes are explanatory of the diagrams:—Plate XXVIII, fig. 1, the Shoulders. shoulder noted at 12h. 13m. Fig. 2, shoulders at 12h. 13m. The angles of these shoulders were sharply defined as I have drawn them, but at 12h. 13m. 30s. there was a quivering from shoulder to shoulder, lasting about a second. I saw both shoulders for 2½ minutes, and when the north one in the inverted image disappeared, the other remained precisely as before for about the same time and then disappeared. Fig. 3, the change from clear contact which I saw to this Quivering. figure was almost instantaneous, a momentary quivering was perceptible, and then this figure was clear and steady and remained so for 3½ seconds, when the dullness next the planet disappeared and was replaced by a clear line of light; it was followed by a still clearer light on the sun's edge, then the black centre disappeared; the whole change scarcely occupying more than one second of time, and at 12h. 23m. 43⅜s. the sun's limb was clear. Fig. 4, from first contact at egress the planet's outline was seen distinctly for 6 minutes and then it disappeared, but a minute later I saw it by the very delicate line shown in this figure. Fig. 5, the slender silvery line, as seen 7½ m. after first contact at egress, lasted, more or less, untill 11m. after that time, when only the first part seen remained, and a minute later this was lost.

H. J. BOLDING,
Police Magistrate.

Messrs. Belfield and Park's Report.

Eversleigh, Armidale, 9 December, 1874.

Weather. — The day was cloudless and free from haze, the air favourable for observation, and at the commencement of the transit definition of the sun's limb clear and sharp.

Instruments. — Instrument used, 4¼-inch refractor, by Cooke, equatorially mounted, no driving-clock; full aperture used, power 130, with first surface reflecting solar eye-piece.

Chronometer showing Sydney mean time nearly and losing rate supposed to be 1·5s daily.

Ingress.

Ingress. — External contact not seen.

At 12h. 14m. 45s., when the planet appeared to be about ⅓ on the sun's disc, the following third of the planet was slightly elongated and its limb distinctly illuminated, giving the appearance of a thin crescent and thus exhibiting the whole disc of the planet most clearly. (Plates VIII and XIX.)

At 12h. 15m. 45s., these appearances becoming more distinct, the air being very good and limb of sun and whole outline of planet being remarkably well defined.

The elongation of following side of planet disappeared as internal contact approached.

Internal contact. No drop. — Internal contact at 12h. 20m. 22s., discs being tangential, no appearance of drop, shade ligament, or other distortion.

At 12h. 20m. 30s., light of sun visible all round the planet, limbs of both bodies still sharply defined and clear.

While Venus was advancing to about ¼ of her own diameter upon the sun, a faint tremulous shading was seen between the edge of the planet and the limb of the sun (both bodies being very sharp in outline), which disappeared so gradually that it could not be said to have been obliterated at any particular instant. (Plates IX and XX.)

Passage. — When fairly on the sun the body of the planet appeared intensely bluish black in centre, becoming of a gorgeous deep blue towards the circumference, which remained well defined and sharp. No appearance of satellite, spots, or nebulous outline. (Plate VIII.)

At 3 p.m., sun boiling, limb of sun and circumference of planet seething, latter losing its blue colour and becoming blacker.

Egress.

Egress. — When Venus approached the sun's limb a shade similar to that observed at ingress was caught, but not so plainly; it in no way interfered with the sharpness of outline of either body. (Plates IX and XX.)

Last internal contact. — At 3h. 52m. 23sec. the shade no longer visible; sun's light still visible all round the planet. At 3h. 53m. 12s., internal contact, edges fairly defined, but not so clear as at ingress. No drop, ligament, or other distortion visible. At 3h. 56m., preceding limb of planet illuminated on one-third of its arc.

Bisection of planet estimated at 4h. 7m. 22s., external contact at 4h. 21m. 51s., definition uncertain, limb of sun running like a mill-race.

Three drawings (Plates X, XIX and XXI), showing northern limb of sun and illuminated edges at ingress and egress.

Plates IX and XX show the tremulous shading just before it disappeared and also the colour on the planet. In the telescope the blue was deep, pure, and utterly beyond practical representation; the same remark applies to the other effects depending upon colour. The tremulous shading is nearly accurate, but there was a quivering motion, like that of heated air on a hot day, which cannot be imitated.

Plate X shows the halo or illumination at egress.

A. W. BELFIELD.
A. J. PARK.

Mr. Belfield's Report—Additional Notes.

My dear Sir,   Evereleigh, Armidale, 24 December, 1874.

I wish to thank you for your letter of the 15th inst., and have much pleasure in replying to your questions and in forwarding coloured sketches which represent what I saw as well as I can transfer my recollections to paper.

The illumination of the following limb of the planet at ingress seemed to be equal in breadth all round, and was bright enough to give an appearance of elongation to the part of the planet not on the sun's disc; it seemed to be outside the limb, which was quite distinct its breadth I see has been estimated at 1"—I should have thought it was nearer 2"—when Venus was ⅔ on the sun's disc, but I have no experience in making measurements of the kind I saw nothing like margin or shading round the part of the planet on the sun's disc.

The blue colour on the planet was the deepest prussian blue, black in centre, lighted towards the edge, but the lightest part was a very deep blue. It did not in the slightest degree mar the clear definition of the planet's disc or extend beyond it, and was not in the least like the violet colour seen round Venus at her brighter phases when viewed against the evening sky. The colour is fairly represented in Plates VIII and IX. There was no trace of colour visible when looking at sun-spots with same diagonal, eye-piece and shades.

The colour was very vivid during the earlier part of the transit, when the air was very steady; during the latter part when the sun began to boil the planet to my eye was a dead black (Plate X); others with me said they saw the blue colour still, to me it was not obvious but had to be looked for; there was certainly a very great difference in colour during the earlier and later parts of the transit. I did not notice the film of cloud you mention.

At egress the illumination was seen on only a part of the planet's disc off the sun. At the intersection of the limbs of Venus and the sun on the northern side of the planet (direct image) it was broader than at any part during ingress, but thinned off to nothing rapidly, and the rest of the planet's outline was not distinguishable from the black background of the sky.

When Venus was about ¼ off the sun the illumination extended over about ½ of the arc of the planet outside the sun, but as the planet's disc left the sun the extent of illumination did not increase accordingly, but was confined to about the same extent of outline as when first noticed. The air at egress was by no means good.

A. W. BELFIELD.

[Forty-one plates.]

Egress at $3^h\ 54^m\ 37^{sec}5$ Sydney mean time.
 Cusps of sunlight separate slightly thickened at extremities and connected by the white light of the ring

Transit of Venus. 1874

Woodford N S W

Mr Vessey's Observations

N

Egress_"Apparent" internal contact at $3^h.54^m.48^{...}41$ Sydney mean time.
Limbs of planet and sun tangential and ring of light slightly overlapping limb of sun

Transit of Venus 1874

Woodford, N S W.

Mr Vessey's Observations

Ingress – showing elongation, illumination and colour

Transit of Venus 1874

N. S. W.

Mr A. H. Belfield's Observations

Ingres — Drawing color of women and snake between the lid

Poster   72   1971

Willson

Egress – showing partial illumination

Transit of Venus 1874

N. S. W.

Mr A. H. Belfield's Observations.

Flash of Light on outer edge of Planet at Egress

Transit of Venus 1874

Mr E. G. Savage's Observations.

Planet elongated

Transit of Venus 1874.

N. S. W.

Mr Watkins Observations.

Black Drop

Transit of Venus 1874

NSW

Venus seen with a narrow fringe of dull red light.

Transit of Venus 1874.

Mr. C.D.Hirst's Observations

Halo on Sun

Transit of Venus 1874.

Cap.ᵗⁿ Onslow's Observations.

Halo and Polar Spot
Transit of Venus, 1874

Halo around planet

Transit of Venus 1874

N S W

M.^r Ma:Donnell's Observations

Shewing positions of illuminated edges

Transit of Venus 1874

Mr A.J.Park's Observations

Figure. Showing illumination of preceding limb. Dark smoky normal observed.

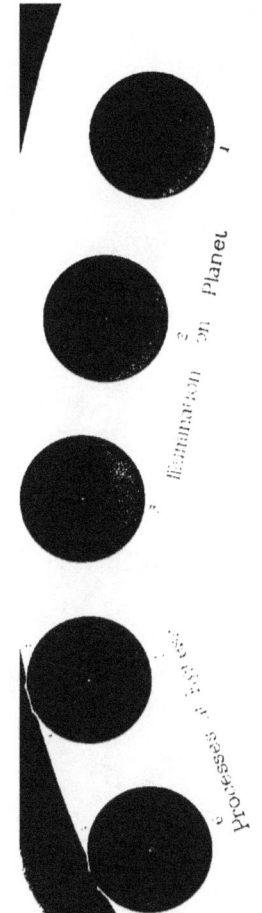

Transit of Venus 1874.

Professor Liversidge's Observations

Halo and Polar Spot seen at egress

Transit of Venus 1874

Mr Russell's Observations

Polar Spot last seen
$4^h\ 23^m\ 22^s$

Polar spot greater haze on Planet.

Halo and Polar spot with haze on Planet
$4^h\ 12^m\ 0^s$

Halo
$3^h\ 57^m\ 7^s$

Transit of Venus 1874

Sydney N.S.W.

Mr Russell's Observations at Egress

Engraved record of illumination curve contact at Ingress and after contact at Egress

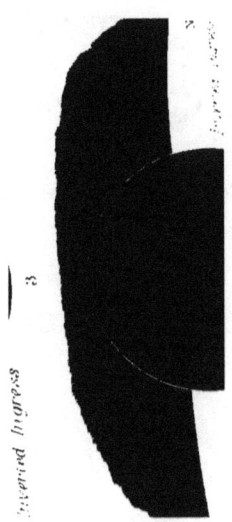

PLATE XXIX.

Transit of Venus 1874.
Sydney, N.S.W.
Mr. A. Fairfax's Observations.

Transit of Venus 1874.
Mr Russell's Observations.

PLATE XXXI.

SYDNEY OBSERVATORY.

PLATE XXXII.

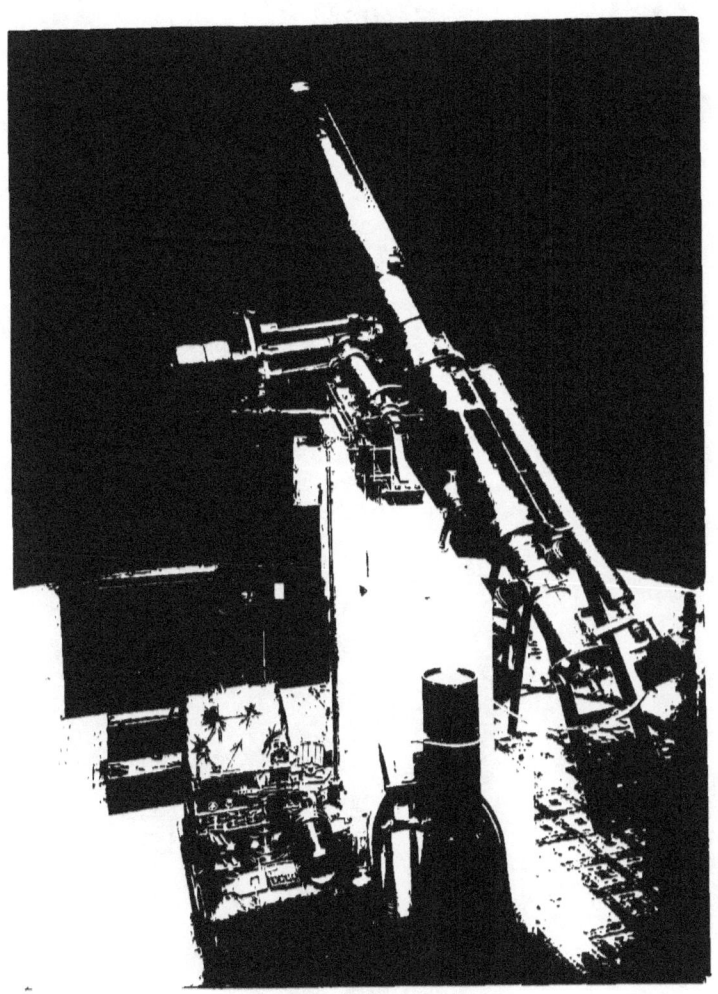

LARGE EQUATORIAL, SYDNEY OBSERVATORY.
OBJECT GLASS 11¼ INCHES.

PLATE XXXIII.

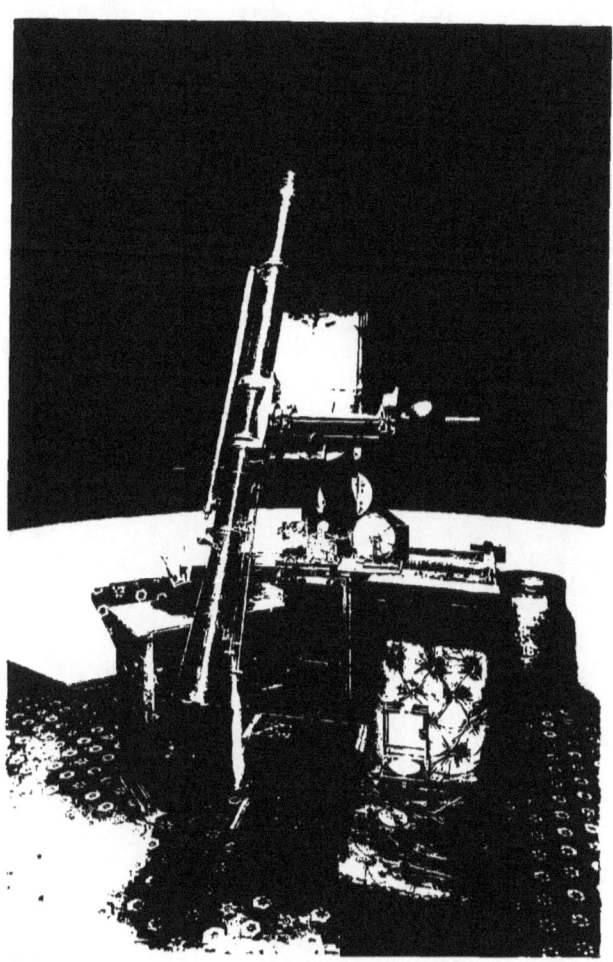

PHOTOHELIOGRAPH,
WOODFORD.

PLATE XXXIV.

TRANSIT OF VENUS CAMP, WOODFORD, NEW SOUTH WALES.

FLATE XXXV.

WAITING FOR THE TRANSIT, AT WOODFORD.

PLATE XXXVI.

WAITING FOR THE TRANSIT OF VENUS, EDEN.

PLATE XXXVII.

7½-INCH EQUATORIAL USED AT EDEN.

PLATE XXXVIII.

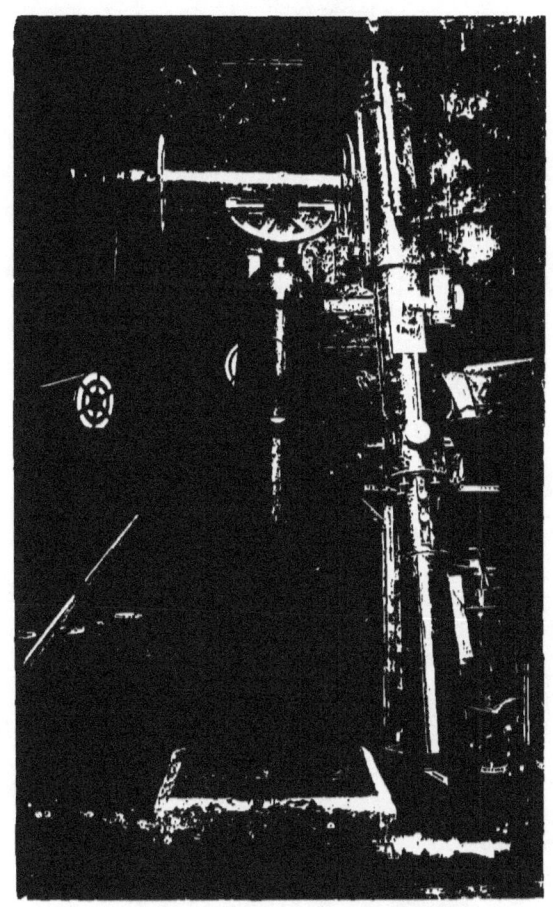

6-INCH EQUATORIAL USED AT GOULBURN.

PLATE XXXIX.

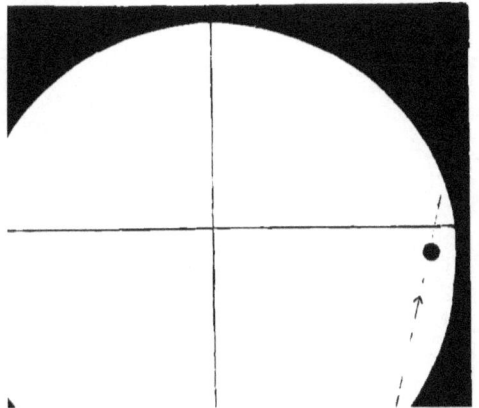

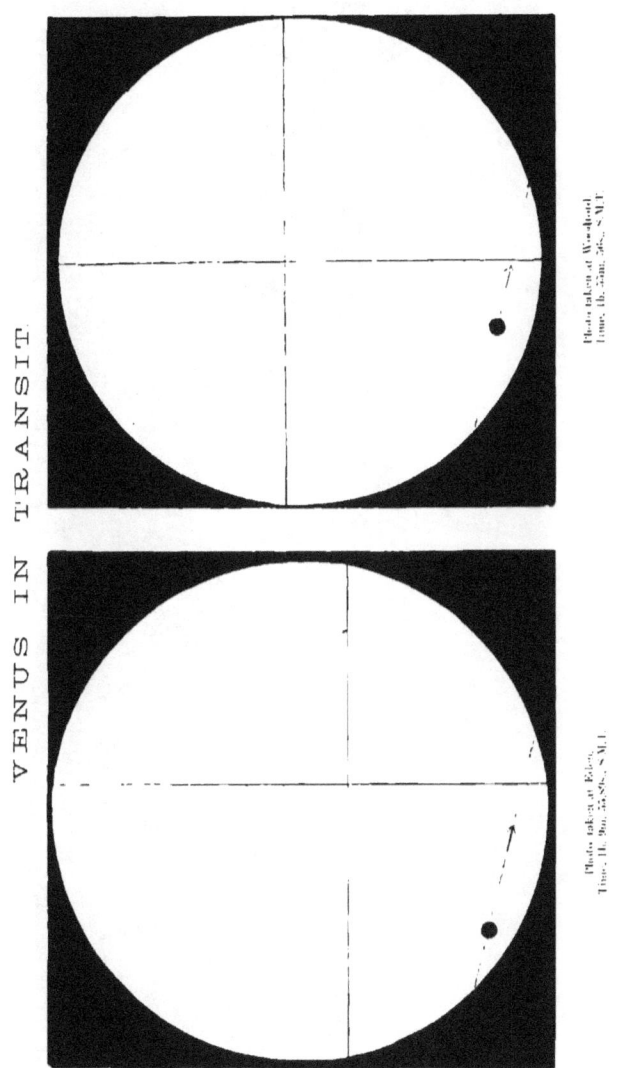

www.ingramcontent.com/pod-product-compliance
Lightning Source LLC
Chambersburg PA
CBHW030317170426
43202CB00009B/1037